# Biogeochemical Transformation in Lake Superior

# Biogeochemical Transformation in Lake Superior:

## *A Radiocarbon and Stable Isotope Study*

By

Prosper Kojo Zigah

**Cambridge Scholars** Publishing

Biogeochemical Transformation in Lake Superior:
A Radiocarbon and Stable Isotope Study

By Prosper Kojo Zigah

This book first published 2025

Cambridge Scholars Publishing

Lady Stephenson Library, Newcastle upon Tyne, NE6 2PA, UK

British Library Cataloguing in Publication Data
A catalogue record for this book is available from the British Library



ISBN: 978-1-0364-1768-0
ISBN (Ebook): 978-1-0364-1769-7

I dedicate this book to my children Ian Landon Zigah and Ivan Carter Zigah. I also dedicate this book to my parents John Zigah (deceased) and Rejoice Adetsi, and to my siblings Patience Zigah and Godwin Zigah.

# CONTENTS

# ACKNOWLEDGEMENTS

I thank my advisors Dr. Elizabeth C. Minor and Dr. Josef P. Werne for giving me the opportunity to work in their laboratories for my doctorate degree. I also thank all my Ph.D committee members, James Cotner, Erik Brown, Richard Axler, and Steve Colman for their excellent guidance, encouragement, and direction. I thank Daniel Repeta and Ann McNichol of Woods Hole Oceanographic Institution for hosting and allowing me to use their laboratories for some of my work. My gratitude goes to the faculty at the Large Lakes Observatory for all their encouragements and support during my doctoral studies. Robert Hecky and S. Leigh McCallister were very resourceful to me throughout this work.

I thank Sarah Grosshuesch and Koushik Dutta and the staff at National Ocean Sciences Accelerator Mass Spectrometry Facility for helping with some of my laboratory analyses. Finally, I thank all my friends from University of Minnesota, University of California Irvine and Woods Hole Oceanographic Institution for their support and fun times during my doctorate studies.

The United States National Science Foundation, National Ocean Sciences Accelerator Mass Spectrometry Facility, Woods Hole Oceanographic Institution and University of Minnesota provided funding for my doctorate research.

**Dr. Prosper K. Zigah**
Department of Chemistry and Biochemistry & Large Lakes Observatory, Water Resources Science Program, University of Minnesota Duluth, Minnesota, USA. **Current address**: PKZ Climate Institute, Georgia, USA.
Email: pkzclimateinstitute.edu@gmail.com
Phone: +1 774 368 0798

**Dr. Elizabeth C. Minor**
Large Lakes Observatory and Department of Chemistry and Biochemistry, University of Minnesota, Duluth, Minnesota, USA.

**Dr. Josef P. Werne**
Department of Geology and Environmental Science, University of Pittsburgh, Pennsylvannia, USA.

**Dr. S. Leigh McCallister**
Department of Biology, Virginia Commonwealth University, Richmond, Virginia.

**Dr. Hussain A.N. Abdulla**
Department of Chemistry & Biochemistry, Old Dominion University, Norfolk, VA 23529, USA. **Current address**: Department of Physical and Environmental Sciences, Texas A&M University, Corpus Christi, Texas, USA.

**Dr. Patrick G. Hatcher**
Department of Chemistry & Biochemistry, Old Dominion University, Norfolk, VA 23529, USA.

**Dr. Ann P. McNichol**
National Ocean Sciences Accelerator Mass Spectrometry Facility, Department of Geology and Geophysics, Woods Hole Oceanographic Institution, Woods Hole, MA, USA.

**Dr. Li Xu**
National Ocean Sciences Accelerator Mass Spectrometry Facility, Department of Geology and Geophysics, Woods Hole Oceanographic Institution, Woods Hole, MA, USA.

# PREFACE

The primary goal of the work presented in this book is to use the distributions of radiocarbon (and also stable carbon and nitrogen isotopes) to determine the sources, ages, and bioreactivity of dissolved organic matter (DOM), particulate organic matter (POM), and dissolved inorganic carbon (DIC) (in this case only sources and ages) in Lake Superior. Specific objectives include:

1. To determine the provenance and biogeochemical cycling of organic matter in the water column of the western arm of Lake Superior using higher temporal sampling resolution. The idea was to capture seasonal and annual dynamics in sources, ages, and turnover of the various carbon pools using their radiocarbon and stable isotopic compositions at the surface productive waters, deep chlorophyll layer, and the deeper hypolimnetic waters.
2. To assess the large-scale (that is the lake-wide) radiocarbon and stable carbon isotopic compositions and the carbon concentrations within DIC, DOC, and POC, and the stable nitrogen isotopic concentrations for DOM and POM, during both stratified and isothermal lake conditions. The focus here was to examine the lateral heterogeneity of the sources and biogeochemistry of organic matter in the lake using carbon concentration and isotopic distributions, and the effect of thermal stratification on the dynamics of mean ages, bioreactivity and origins of organic carbon across the lake.
3. To examine the radiocarbon and stable isotopic distributions within specific physical size-fractions of DOM, and the Nuclear magnetic resonance (NMR) spectra of high molecular weight (HMW) DOM, to more robustly constrain the sources and cycling of DOM in Lake Superior. DOM is the largest organic carbon pool in the lake, and its dynamics have implications for biogeochemistry of the lake ecosystem as well as regional climate. Finer evaluation of the sources and turnover rates of the constituent size fractions of DOM is therefore fundamental to our understanding of OM dynamics in the lake.
4. To investigate the radiocarbon and stable carbon isotopic composition of organic fractions of high molecular DOM in the lake. As with multiple sizes, DOM also contains biochemical fractions with numerous sources, reactivity, and different formation and diagenetic pathways. A

better constraint on this considerably enhances the understanding DOM biogeochemistry and roles in the lake.

5. To examine the basal food sources sustaining zooplankton in Lake Superior using natural abundance radiocarbon and stable carbon and nitrogen distributions. This ultimately will help in understanding the fate of terrestrial and old (sediment-derived) OM in the lake, that is whether it is channeled up the food web, or not, and will also provide insights into which food options are potentially critical in sustaining economically important fish in the lake.

The organization of the book is presented below.

CHAPTER ONE gives introductory overview of the sources and transformation of total or bulk organic matter and its constituent biochemical fractions, and size-fractionated organic matter. It presents the use of radiocarbon as a time and source tracer. And also, spectroscopic insights into the sources and recycling of organic matter in aquatic systems.

CHAPTER TWO addresses the first objective by evaluating the sources and cycling of DIC, DOC, and POC in the western water column of Lake Superior using multiyear distributions of radiocarbon and stable carbon isotopic compositions. Samples were taken during spring mixing and late summer thermal stratification over a two-year period (2007-2009).

CHAPTER THREE presents lake-wide distributions of the radiocarbon and stable carbon and nitrogen isotopes to gain a large-scale picture of the sources and transformations of OM in the lake. Data was collected from 8 stations across the lake sampled during the isothermal and thermally stratified lake water column.

CHAPTER FOUR examines the carbon isotopes ($\Delta^{14}C$ and $\delta^{13}C$) and concentrations of four size fractions of DOM, and the proton and $^{13}C$-NMR of HMW DOM to further investigate DOM dynamics and sources.

CHAPTER FIVE presents the natural abundance carbon isotopes ($\Delta^{14}C$ and $\delta^{13}C$) of bulk and thermal fractions of solid phase extracted (SPE) DOC and high molecular weight (HMW) DOC. It also examines the organic fractions within HMW DOC including extractable lipids, hydrolyzable carbohydrates, amino acids, and acid insoluble organic fraction (which is part of molecularly uncharacterized fraction) in order to investigate the sources, turnover times, and cycling of DOC in Lake Superior.

CHAPTER SIX presents radiocarbon ($\Delta^{14}C$) and stable isotope ($\delta^{13}C$, $\delta^{15}N$) analyses to show that mesozooplankton in Lake Superior selectively feed on within-lake produced OM even though allochthonous carbon sources represented a considerable portion of the available food resource.

CHAPTER SEVEN is a summary of the conclusions of the studies presented and recommendations for future studies.

CHAPTER EIGHT is the list of references.

# ABOUT THE AUTHOR

Dr. Zigah is the President and CEO of PKZ Climate Institute. Before then, Dr. Zigah was a visiting Assistant Professor of Chemistry in the Department of Chemistry, Biochemistry and Physics at Georgia Southern University. Professor Dr. Prosper Zigah is a scientist and a teacher. He holds professional certificates from Harvard University and University of Cambridge. He completed his Doctor of Philosophy degree (PhD) in Water Resources Science at the University of Minnesota. He holds two master's degrees in Environmental Science from Wageningen University and the University of Nottingham. He got his bachelor's degree in Agricultural Science from the University of Cape Coast.

Dr Zigah has held research scientist positions at Swiss Federal Institute of Aquatic Science and Technology, Woods Hole Oceanographic Institution and the University of Pittsburgh.

Dr. Zigah has been a visiting researcher at Harvard University, University of Massachusetts Amherst, National Ocean Sciences Accelerator Mass Spectrometry Facility, University of Florida and the International Water Management Institute.

Dr. Zigah has two lovely boys Ian Landon Zigah and Ivan Carter Zigah. He has lived and worked in several US states including Florida, Maryland, Minnesota, Massachusetts, Pennsylvania and Georgia. He has also lived and worked in the Netherlands, United Kingdom and Switzerland. In his free time, Dr. Zigah enjoys hiking, jogging, traveling, cooking, watching soccer, watching football, and writing.

# CHAPTER ONE

# INTRODUCTION

## 1.1 Sources and biogeochemical transformations of organic matter in aquatic systems

Organic matter (OM) is a major reactive component in aquatic systems and is ultimately derived from in situ primary production (autochthonous) or from terrestrial plant debris (allochthonous). The relative importance of these sources varies, and is related to surface area, size of drainage basin, nutrient availability, and trophic status.

Measurements of organic carbon are usually used to determine the relative amounts of organic matter present in a system. In most aquatic systems, especially the oceans and large lakes (surface area > 10, 000 $m^2$), suspended particulate organic carbon (POC), usually measured on particulate organic matter collected on filters (POM) is usually a much smaller pool than dissolved organic carbon (DOC), albeit a more dynamic one and a major conduit for coupling surface water and deep water biogeochemical processes such as photosynthesis and respirations, and surface-to-sediment transport of OM (Mcknight et al. 1997; Druffel and Bauer 2000; Volkman and Tanoue 2002; Verdugo et al. 2004; McNichol and Aluwihare 2007). However, in riverine and estuarine systems, the POC pool can be comparable in magnitude to DOC (Richey et al. 1990; Fisher et al. 1998). Most studies of POM over the past several decades have been conducted to understand its fluxes, biogeochemical transformations, and fate in aquatic systems. These studies, most of which were in oceanic systems, have shed light on the transit times of sinking POM from the surface to deep ocean, and also helped in quantifying what fraction of POC produced in the surface waters is able to evade oxidation/remineralization and get sequestered in the sediments (Bacon and Anderson 1982; Lee et al. 2004; McNichol and Aluwihare 2007). Sequestration of organic matter in sediments is an important control of atmospheric carbon dioxide ($CO_2$) levels. A major fraction of POM, which becomes a larger proportion of total remaining POM in deeper waters and surface sediments, remains molecularly uncharacterized (MUC); understanding this MUC continues to be an active field of research (Wakeham et al. 1997; Hedges et al. 2000; Hwang and Druffel 2003; Lee et al. 2004; Hwang and Druffel 2006).

Dissolved organic matter (DOM), operationally defined as the fraction of OM that passes through the 0.1-1.0 μm filter, consists of a heterogeneous mixture of biomolecular and geochemical compounds. The importance of DOM in the global carbon cycle is reflected in the fact that DOC is the largest OC pool (~90%) in the oceans and large freshwater systems, and one of the Earth's largest reactive carbon pools. The magnitude of DOC in the world's oceans (~685 Gt C) is similar to the magnitude of atmospheric $CO_2$ (Hansell and Carlson 1998; Hedges 1992). Accordingly, perturbations of DOC in terms of varying pool sizes could have considerable implications for the global biogeochemical systems. Therefore DOC fate in aquatic systems has potential implications for local, regional, and global climate dynamics if there are shifts in the rate or extent of DOC mineralization to atmospheric $CO_2$. For example, additional mineralization of just 1% of oceanic DOC would generate more atmospheric $CO_2$ than annual fossil fuel emissions (Hedges 2002; Mopper et al. 2007). DOC also plays key roles in nutrient recycling and aquatic chemistry: serving as food resource for heterotrophic bacteria and zooplankton, attenuating harmful UV-B radiation and protecting DNA and pigments from damage, pH buffering, regulating toxicity and bioavailability of pollutants and trace metals (Hedges 1992; Balakrishna et al. 2006; Klaminder et al. 2006).

Despite being one of the largest reservoirs of carbon on earth with key physicochemical and biogeochemical roles, the fate of DOC in aquatic systems is poorly constrained. Our understanding of the sources and cycling of DOC is complicated by the numerous sources, heterogeneous size and molecular distributions, and varying diagenetic transformations and biogeochemical sinks (Hedges 1992; Guo et al. 1996; McCallister et al. 2006; Beaupre and Druffel 2009). For instance, while several studies have reported the transport of terrestrial OM to oceans (~0.25 Gt DOC per year) (Opsahl and Benner 1997; Raymond and Bauer 2001; Cauwet 2002), to rivers (Saliot et al. 2001; Kaiser et al. 2004), and to lakes (Cole et al. 1994), the fate of the terrigenous OM in these aquatic systems are not well known. In the oceans, while Hedges et al. (1997) suggested that estuarine mixing processes account for the rapid degradation of terrigenous OM en route to the oceans, other studies indicate a lesser role for this process in removing terrigenous OM (Doering et al. 1994). Also, whereas Moran et al. (1991) noted that microbial degradation in the ocean margins was the main mechanism that degrades terrigenous OM, Benner and Opsahl (2001) in their study in the Mississippi river plume indicated that photochemical oxidation was rather the main sink for terrigenous OM. In lakes, although the terrigenous OM subsidy accounts for a generally net heterotrophic nature (del Giorgio and Peters 1994; Kritzberg et al. 2005; Karlsson et al. 2007), the terrestrial OM flux and transformation pathways are not well constrained.

OM diagenetic transformation and cycling in aquatic systems is mainly mediated by microbial degradation and/or photochemical oxidation as noted by studies in rivers (Benner et al. 1995; Opsahl and Benner 1998), estuaries (Benner and Opsahl 2001; Wang et al. 2004), oceans (Miller and Zepp 1995; Miller and Moran 1997; Opsahl and Benner 1998) and lakes (McCallister et al. 2004). At surface-to-deep water integrated level, microbial mineralization is generally thought to be the dominant OM oxidation process (Cotner et al. 2004), and its extent is related to oxygen level, ionic strength, pH and temperature (Benner 2004). Photochemical oxidation of DOM occurs in the mixed surface waters where the DOM absorbs ultraviolet radiation, degrading the humic substances and other large colored dissolved organic compounds into smaller molecules (Moran and Zepp 1997). Studies show that photo-oxidation makes OM either more biolabile (through nutrient release or alteration of structure to a more edible form) or less biolabile (due to photochemical crosslinking and other OM alterations as well as production of radical oxygen species such as singlet oxygen, superoxide, peroxides and hydroxyl radical (Cotner et al. 2004)). Other processes such as enzymatic hydrolysis (Arnosti and Holmer 2003), flocculation (Fox 1983; Wang et al. 2004), sedimentation and solubilization (Hedges 1992; Prahl et al. 1994; Azam and Long 2001) and physical mixing and transport (Findlay et al. 1998; Cloern et al. 2002) could modify and/or recycle OM in aquatic systems.

Due to the numerous sources and fates of OM as noted earlier, constraining its distributions and processing is challenging, albeit fundamental to understanding the carbon sources supporting the aquatic food web. The fate of the autochthonous and/or terrestrially-derived OM in aquatic systems, and whether these OM components are permanently sequestered in the sediment, channeled to higher trophic levels in the food web or remineralized to $CO_2$, is critical to understanding the timescales of the atmospheric $CO_2$ sink, and/or generation of atmospheric $CO_2$ and to evaluating the food sources for economically important fishes. Therefore, the fate of OM in aquatic systems could potentially have implications for regional and global climate dynamics through heat exchange and effects on radiative forcing and could impact ecosystem function (including the global fishery) as OM is the source of energy and biomass for heterotrophs.

Natural abundance stable isotopes have proven useful for identifying sources and biogeochemical processes occurring in aquatic environments (Hedges et al. 1988; Prahl et al. 1994). Stable carbon isotope ($\delta^{13}C$) values have been utilized as a tool to estimate relative OM from autochthonous and allochthonous sources based on the general $^{13}C$ depletion in terrestrial organic materials. However because of the smaller dynamic range of OM $\delta^{13}C$ values (-32 to -12‰, McCallister et al. 2004; Wakeham et al. 2006), and the considerable overlapping of $\delta^{13}C$ values of numerous OM sources, its use for inferring OM sources in aquatic systems is usually difficult. The natural abundance radiocarbon ($\Delta^{14}C$) signature in OM samples can act as a robust time, source and process tracer because of its larger dynamic range (-1000 to ~+200‰) (Petsch et al. 2001; McCallister et al. 2004; Wakeham et al. 2006). The isotopic, biomolecular, and chemical compositions of OM have been successfully utilized as proxies for identifying sources and biogeochemical processes occurring in aquatic environments (Hedges et al. 1986; Raymond and Bauer 2001; Hwang et al. 2004; Williams et al. 1992; Druffel et al. 1992). Each of these is described in detail in the subsequent sections (see sections 1.2 to 1.5). The main focus of this work is to use natural abundance radiocarbon, coupled to an additional suite of parameters ($\delta^{13}C$, $\delta^{15}N$, C:N ratio) within bulk, and size and compound fractions of OM to assess the sources, relative ages, and the potential bio-reactivity of POC and DOC (and to compare with the dissolved inorganic carbon or DIC pool) in Lake Superior, the Earth's largest freshwater lake by surface area.

## 1.2 Radiocarbon production, and its use as a time, source and process tracer

Radiocarbon is naturally generated in the upper stratosphere and lower troposphere by highly energetic charged particles (cosmic rays) colliding with molecules of the earth's atmosphere generating neutrons, which then interact with nitrogen to produce $^{14}C$ (McNichol and Aluwihare 2007). The $^{14}C$ quickly combines with oxygen into $CO_2$ and becomes incorporated into the terrestrial biosphere through photosynthesis. Radiocarbon declines with a half-life of 5730 years making it amenable for studying surficial environmental processes over the past 60,000 years

(Petsch et al. 2001). The use of $^{14}C$ for determining the age of organic materials is constrained by fluctuations in the flux of cosmic rays, dilution effect from combustion of fossil fuels and the $^{14}C$ spike in the atmosphere caused by above-ground thermonuclear weapons testing during the 1950-1960s. The bomb spike artificially elevated the $\Delta^{14}C$ values of $CO_2$ from ~ -20‰ in the pre-1950 atmospheric air to ~+1000‰ in 1964 at the peak of the above-ground bomb testing. Similarly, $\Delta^{14}C$ of oceanic surface dissolved inorganic carbon (DIC) was elevated from -50‰ before 1950 to approximately +110‰ in 1964 (Nydal 1963; Nydal et al. 1980; William et al. 1992; Bauer et al. 2001; McNichol and Aluwihare 2007).

As a result of the bomb spike, organic matter synthesized from inorganic carbon in equilibrium with atmospheric $CO_2$ has elevated $^{14}C$ content. This $^{14}C$ elevation has provided opportunity for determining the importance of modern (post 1950s or post bomb) and pre-aged (pre-1950 or pre-bomb) organic carbon in the biogeochemistry of aquatic systems, including Lake Superior. However, the atmospheric $^{14}C$ spike has been steadily declining after the 1964 (McNichol and Aluwihare 2007) due to cessation of the above-ground testing of nuclear devices and the equilibration of the excess atmospheric $^{14}C$ into the carbon cycle. This can lead to the potentially confusing situation where material photosynthesized 20 to 30 years ago will have a more enriched signature (i.e., reflect a more "modern" signal) than material being photosynthesized today. However, on decadal time-scales the approach can be applied to study ancient (pre-bomb) versus contemporary (post-bomb) material.

The natural abundance of $^{14}C$ within the DIC, POC and DOC reservoirs allows for the determination of their relative ages and the various processes that control their dynamics in the lake ecosystem. The $\Delta^{14}C$ (per mil deviation of $^{14}C/^{12}C$ ratio relative to nineteenth century wood standard) contents of carbon pools have been used by several researchers for studying carbon sources and cycling in rivers (Hedges et al. 1986; Raymond and Bauer 2001) and oceanic systems (Druffel et al. 1992; Hwang et al. 2004; Williams et al. 1992; Nagao et al. 2005). Despite the successes of these studies, $\Delta^{14}C$ has not been used to comprehensively examine the sources and biogeochemical sinks of OM in any of the great lakes of the world, including the Laurentian Great Lakes of North America.

## 1.3 Sources and cycling of size-fractionated and compound-class-fractionated DOC based on natural abundance radiocarbon composition

The $^{14}C$ signals of fraction-specific DOC such as colloidal or high molecular weight (HMW) DOC (> 1000 Da) and low molecular weight (LMW) DOC (< 1000 Da) have gained currency as a tracer of OC source and fate in aquatic systems (Guo et al. 1996; McNichol and Aluwihare 2007). These measurements, however, allow only a small fraction (< 20-40%) of the DOC reservoir to be studied as compared to the total (bulk) DOC pool (Benner et al. 1992; Guo et al. 1994; McNichol and Aluwihare 2007). Several studies have used the cross-flow ultrafiltration technique for isolating and concentrating large amounts of HMW DOC for radiocarbon and biomolecular analyses in order to investigate the provenance and reactivity of DOC (Bianchi et al. 1995; Guo et al. 1996; Loh et al. 2004). HMW DOC is usually enriched in contemporary ($^{14}C$-enriched) labile biochemicals relative to the concurrent bulk DOC and recycles rapidly. The LMW DOC on the other hand is typically depleted in $^{14}C$ as compared to the bulk DOC and recycles on a much longer time scale (Guo et al. 1995; Santschi et al. 1995; Guo et al. 1996). In their study in the Mid-Atlantic Bight (MAB), Aluwihare et al. (2002) noted that in the surface and deep waters, HMW DOC was more $^{14}C$-enriched than the bulk DOC by 22‰ and 150‰, respectively. Accordingly, HWM DOC represents the modern bio-reactive fraction of the bulk DOC whereas refractory organic molecules tend to concentrate in the LMW DOC. Different DOC size fractions could have unique origins, and be subject to varying diagenetic transformations. Therefore, it is a critical first step to characterize the amount, age, and biochemical nature of various DOC size classes in addition to the bulk DOC.

In addition to size fractions, the bulk DOC in aquatic systems can be split into different organic fractions that could have unique provenances and diagenetic transformations as well. The $^{14}C$ signatures of organic compound classes such as carbohydrates, lipids and proteins are important in teasing apart specific sources and the turnover times of molecules and compounds within the bulk DOC. Typically, however, a large amount of organic material has to be gleaned and concentrated in order for their constituent biochemicals to be isolated. Because size fractionation into HMW DOC and LWM DOC does not alter the chemical composition of the isolates and has been shown to have a low carbon blank, most compound class radiocarbon investigations are carried out on HMW DOC

(Santschi et al. 1995; Aluwihare et al. 2002; Loh et al. 2004; Wang et al. 2004; Guo and Macdonald 2006; Guo et al. 2009; Walker et al. 2011). Previous studies in various aquatic systems show that carbohydrates are the most abundant biochemicals within the HMW DOM, and are, accordingly, the most common macromolecules that have been $^{14}C$ characterized (Wang et al. 1998; Santschi et al. 1998; Loh et al. 2004; Aluwihare et al. 2002; Repeta and Aluwihare 2006). Studies in oceanic systems show that carbohydrates are typically more $^{14}C$-enriched than concurrent HMW DOM. For instance, the $\Delta^{14}C$ of carbohydrates in the surface waters of Mid-Atlantic Bight (Santschi et al. 1998), North Atlantic, and North Pacific oceans (Loh et al. 2004) were 138‰, 18‰, and 99‰ more $^{14}C$-enriched than the co-occurring HMW DOM. Repeta and Aluwihare (2006) used radiocarbon to study neutral sugars in the North Pacific Ocean, and noted that carbohydrates cycles faster with a residence time of $\leq$ 3 years in the surface ocean. Similarly, proteins recycle faster and are typically $^{14}C$-enriched relative to concurrent HMW DOC at all depths. In their study in the open Pacific and Atlantic Oceans, Loh et al. (2004) reported that protein-like substances in the surface waters were respectively 7‰ and 71‰ more enriched in $\Delta^{14}C$ than concurrent HMW DOM.

Lipids usually make up only a small fraction of HMW DOM as compared to carbohydrates (Loh et al. 2004; Wang et al. 2004). Refractory lipid compounds such as n-alkanes, alkenones, and alcohols tend to be well-preserved in aquatic systems and their distributions have been used successfully as source and/or process-specific identifiers (Meyers 1997; Pearson et al. 2000; Pearson et al. 2001; Druffel et al. 2010). In contrast to carbohydrates and proteins, lipids are usually more $^{14}C$-depleted relative to concurrent HMW DOM and total DOC due to their general refractory nature and long residence times in aquatic systems. In the North Atlantic Ocean for instance, the $\Delta^{14}C$ of solvent extractable lipids was -637‰ (compared to $\Delta^{14}C$ of HMW DOC of -5‰) in the surface waters, and -730‰ (compared to $\Delta^{14}C$ of HMW DOC of -270‰) in the deep waters (Loh et al. 2004).

## 1.4 Spectroscopic insights into the provenance and cycling of organic matter in aquatic systems

Dissolved organic matters have distinct chemical compositions, which reflect their origin and diagenetic state in an aquatic system. In order to obtain enough organic material to study chemical and molecular characteristics and/or properties of OM via compound-class isotopic and nuclear magnetic resonance (NMR) analyses, isolation is generally necessary for concentrating material. There are several methods used for isolating and concentrating DOM depending on the desired isolate needed as some of these do alter and/or isolate specific chemical constituents of the bulk DOM. Some of the common DOM isolation approaches are briefly described here.

Ultrafiltration isolates DOM based on physical size distributions into HMW DOM and LMW DOM (see also Section 1.3), and has been used to characterize the chemical compositions and sources and processing of DOM from several aquatic environments (Minor et al. 2002; Benner et al. 1992; Hernes and Benner 2002; Helms et al. 2008). Also, ultrafiltration has been shown to have negligible blank or contamination issues (Guo and Santschi 1996). Other techniques that are used for isolating and concentrating OM are solid phase extraction techniques such as the use of $C_{18}$ membranes, alkyl chains with 18 carbons that are covalently bonded to a silicate surface (Kim et al. 2003; Simjouw et al. 2005, Minor and Stephens 2008), and XAD resins, stationary phase is made of polystyrene resin (Hedges et al. 1992). In contrast to ultrafiltration, these techniques often employ large pH changes within the sample to maximize separation and isolation of the DOM, and also chemically fractionate the bulk DOM into polar and non-polar fractions, which could be shifted in terms of sources and origins relative to the bulk material (Simjouw et al. 2005; Schwede-Thomas et al. 2005; Kruger et al. 2011).

Ultraviolet (UV)-visible spectroscopy is an analytical approach for studying DOM properties, and it is applicable to whole water samples as well as extracts and HMW DOM and can thus give an idea of the bulk chromophoric DOM as well as how isolation techniques might fractionate samples. The proportion of DOM that absorbs UV-visible light is termed chromophoric or colored dissolved organic matter (CDOM). The UV-visible absorption spectra of CDOM have been used successfully to distinguish between terrigenous and aquatic OM. The ratio of absorption at 250 nm to 365 nm (termed the $E_2/E_3$ ratio) is inversely related to molecular size, and has been used by some researchers to trace the changes in the size of DOM in aquatic systems (Peuravouri and Pihlaja 1997; Minor and Stephens 2008). The UV absorption at 254 nm normalized to DOC concentration (called $SUVA_{254}$) has been shown to correlate strongly with the aromaticity of DOM (Weishaar et al. 2003; Minor et al. 2006), and is therefore a good proxy for terrestrial input. The spectral slope (S, $nm^{-1}$) is computed from the absorption data from the equation: $\alpha_\lambda = \alpha\lambda_{ref}\, e^{-s\,(\lambda - \lambda ref)}$, where $\alpha$ = Napierian absorption coefficient in $m^{-1}$, $\lambda$ = wavelength in nm, and $\lambda_{ref}$ = reference wavelength in nm (Helms et al. 2008). This index is inversely related to molecular weight, and provides insights into the photochemistry, source and cycling of CDOM (Helms et al. 2008).

Nuclear magnetic resonance (NMR) spectroscopy is another powerful analytical approach for studying the broad overall chemical composition of DOM. NMR is non-destructive and allows measurement of the distributions of the major functional groups in DOM. The NMR technique works by determining the chemical environments around nuclei with magnetic moments by looking at the magnetic properties of these nuclei (e.g., $^{13}C$, $^{1}H$, $^{15}N$, $^{31}P$) when placed in strong magnetic fields (Mopper et al. 2007). The peaks or resonances in specific chemical shifts in the NMR spectra could indicate the presence of specific chemical groups or biochemical classes such as carbohydrates, proteins, and lignin (Repeta et al. 2002; Baldock et al. 2004; Abdulla et al. 2010; Gogou and Repeta 2010). NMR spectroscopy has been used to study DOM sources and cycling in several aquatic systems (Malcolm 1990; Benner et al. 1992, Hedges et al. 1992; Aluwihare et al. 2002; Repeta et al. 2002; Bianchi et al. 2004; Abdulla et al. 2010; Gogou and Repeta 2010) based on relative composition and variations in carbohydrates, proteins, aliphatic and aromatic components.

# CHAPTER TWO

# RADIOCARBON AND STABLE CARBON ISOTOPIC INSIGHTS INTO PROVENANCE AND CYCLING OF CARBON IN LAKE SUPERIOR

## 2.1 Introduction

The biogeochemical cycling of carbon in aquatic ecosystems has gained increased attention in the past few decades due to global climate changes encompassing increases in atmospheric $CO_2$, potential hydrologic accelerations and additional attendant ecosystem changes. Due to their proximity to terrestrial environments and the associated delivery of allochthonous organic carbon, inland aquatic systems can function simultaneously as both sinks and sources of atmospheric $CO_2$ (Cole et al. 2007). Multiple biogeochemical processes such as carbonate buffering, heterotrophic respiration and sedimentary sequestration of organic carbon (OC) allow inland aquatic systems such as Lake Superior to outgas $CO_2$ to the atmosphere while concurrently burying OC at depth (Kritzberg et al. 2005; Urban et al. 2005). The bioavailability and cycling of carbon is also important in within-ecosystem processes as inorganic carbon is a basic building block for autotrophy and labile OC provides necessary carbon and energy sources for heterotrophy.

The in situ biological and chemical transformations of OC in aquatic systems are largely dependent on its source, physical packaging (e.g., mineral associations) and prior diagenetic alterations in both the water column and surface sediments (Kaiser et al. 2004). However, the sources and processing of OC are poorly constrained in many ocean, river and lake systems even though delineating these is critical in understanding their transformation pathways and ultimate fate. For instance, even though considerable evidence exists for transport of terrestrial OC to the oceans (Opsahl and Benner 1997), rivers (Raymond and Cole 2003), and lakes (Cole and Caraco 2001), the fate of terrestrial OC in these ecosystems is not well known.

Natural abundance stable and/or radiocarbon isotopes have been useful for studying organic carbon sources and carbon cycling in the oceans (Druffel et al. 1992; Bauer et al. 2002), estuaries (Peterson et al. 1994), rivers (Kaiser et al. 2004; Raymond and Bauer 2001), and lakes (Karlsson et al. 2007; McCallister et al. 2008). The decadal scale resolution added by the spike in $\Delta^{14}C$ resulting from 1950s above-ground nuclear testing has been employed to study C cycling and turnover in both terrestrial and aquatic environments (McCallister et al. 2004; Trumbore 2009) and has provided an additional timescale to that offered by radioactive decay methods. While such studies of 'modern' processes in aquatic ecosystems have yielded insight into C transformation and cycling, to date there has not been a radiocarbon study investigating carbon cycling in any of the great lakes of the world, including the Laurentian Great Lakes of North America (Cotner et al. 2004). Therefore this study (a time series from an open-lake station and a nearshore site in Lake Superior) was designed to assess sources, ages and relative reactivity of particulate and dissolved OC (POC and DOC, respectively) in a temperate and relatively pristine large lake and to compare the stable- and radio- carbon isotopic composition of lake OC with that of co-occurring dissolved inorganic carbon (DIC).

Lake Superior is the Earth's largest freshwater lake by surface area ($8.2 \times 10^{10}$ $m^2$) with maximum and mean depths of, respectively, 406 m and 150 m (Urban et al. 2005). The lake is biogeochemically similar to open-ocean locations due to its oligotrophic nature, low terrestrial nutrient loading and the dominance of its microbial food web on carbon cycling (Cotner et al. 2004). However, unlike the oceans, Lake Superior is dimictic, thus there is density-driven complete vertical mixing of the water column in spring and early winter each year which homogenizes the water column. The dimictic nature of Lake Superior provides a unique opportunity to study how organic matter (OM) from different allochthonous and autochthonous sources can vary in processing and reactivity on short time scales (annual to decadal), an aspect of OM dynamics that is not amenable to in situ study in the oceans since exchange of OM between surface and deep waters generally occurs over very long time scales (McNichol and Aluwihare 2007).

## 2.2 Methods

### 2.2.1 Sampling

Multiple cruises were undertaken on the R/V *Blue Heron* to sample the thermally stratified water column in August or September 2007, 2008,

2009 and the isothermal (mixed) water column in May or June 2008, 2009. Water samples were collected from an open-lake (OL) site (47°19.20'N, 89°49.49'W), chosen to take advantage of data from a National Oceanic and Atmospheric Administration (NOAA) buoy (National Data Buoy Center (NDBC) station 45006) and sediment trap moorings within 1-2 km of the site, and a nearshore site (site Baptism River (BR), 47°19.95'N, 91°11.51'W) just offshore of the Baptism River in the western arm of Lake Superior (Fig. 2-1). At the offshore site (total water depth approximately 165 m), three water depths were sampled each time: the surface (5 m), the depth of the summer deep chlorophyll maximum (DCM) (30 m), and the deep hypolimnion (127 m). The nearshore site (total water depth ~21 m) was sampled only at 5 m.

Water samples were collected via Niskin bottles mounted on a Seabird model 911 plus conductivity, temperature, and depth (CTD) rosette equipped with fluorometer, transmissometer, dissolved oxygen sensor, photosynthetically active radiation (PAR) sensor, pH meter, and altimeter. DIC samples were taken directly from the Niskin bottles by rinsing three times with sample and then overflowing two volumes of the unfiltered water into previously acid-cleaned and combusted (450°C for 4 hours) 500 mL amber Pyrex bottles. Note that these are technically total inorganic carbon samples but Lake Superior water does not contain measurable particulate inorganic carbon. After removing a known and consistent headspace, the samples were immediately preserved with saturated mercuric chloride solution, sealed air-tight with glass stoppers coated with Apiezon grease, and stored at room temperature in the dark.

Water samples for POC and DOC were filtered through pre-combusted Whatman GF/F glass fiber filters (450°C for 4 hours; 0.7 μm nominal pore size) via nitrogen- pressurized stainless-steel canisters. For each DOC sample for radiocarbon analysis, approximately 1 L of the resulting filtrate was collected into an acid-leached and combusted glass bottle. For each DOC concentration measurement, approximately 40 mL of the filtrate was collected into an amber glass vial (previously acid-leached and combusted). Both types of DOC samples were preserved by acidifying to pH 2 (using 6 mol $L^{-1}$ HCl, American Chemical Society (ACS) Plus grade) and stored refrigerated. After ~10 L of lake water had passed through a GF/F filter, the filter and retained particulate matter was removed from the stainless steel holder, folded, placed in previously-combusted aluminum foil and stored frozen until analysis.

To ensure that our sampling techniques are appropriately clean, a large-volume ultra pure water (Millipore Milli-Q Plus) blank was processed by filtering through the canister set-up. This large volume blank consisted of > 100 L of Milli-Q water (note that this is ten times the volume of lake water filtered for POC samples) and was designed to provide enough carbon for a radiocarbon measurement.

Sediment cores from the open-lake site were taken in October 2008, and June and August 2009 using an Ocean Instruments multi-corer. Upon recovery, the overlying water was collected via acid-cleaned syringe to just above the flocculant layer and sediments (with the flocculant layer included in the first section) were sectioned at 2-cm resolution with the depth 'slices' placed in pre-combusted glass jars and stored refrigerated for the duration of the cruise. Upon return to shore, within 9 days of coring, the overlying water DOC and sedimentary pore-water DOC from each depth slice were obtained by centrifuging and filtering the supernatant through pre-combusted glass fiber filters. These pore-water DOC samples were then acidified to pH 2 and stored refrigerated until analysis.

Mesozooplankton samples were collected at site OL via vertical net tows (using a 300 μm mesh net) from 50 m to the water surface. Sampling was done at night, thus taking advantage of zooplankton diurnal migration patterns and maximizing biomass collection for radiocarbon analysis. The biomass was rinsed with lake water into the cod end of the net and concentrated onto glass fiber filters (pre-combusted GF/F filters, 0.7 μm pore size), which were placed in combusted aluminum foil and stored frozen.

Amity Creek was sampled just above its confluence with the Lester River in June and September 2008 during stormflow and baseflow conditions, respectively. Amity Creek drains a small, primarily forested watershed in the western arm of Lake Superior. It flows into the Lester River less than 0.5 km before that river enters western Lake Superior, and is conveniently located near the Large Lakes Observatory. Baseflow samples were taken as grab samples using acid-cleaned carboys while storm flow samples were taken using a Sigma 900 autosampler holding 24 acid-cleaned sample containers, with sampling triggered by stage height changes as determined by a pressure sensor. Whole water samples were removed from the autosampler within 24 hours of the storm event. Both baseflow and stormflow samples were filtered through previously-combusted GF/F filters to isolate POC vs. DOC. The baseflow DIC sample was collected by overflowing two volumes of the unfiltered water into previously acid-cleaned and combusted amber Pyrex bottles. After removal of a known headspace, the sample was preserved with saturated mercuric chloride solution, air-tight sealed and stored at room temperature in the dark until analysis.

In September 2009, corn leaves (*Zea mays*) were collected from the watershed of western Lake Superior in order to determine the radiocarbon content of atmospheric $CO_2$. The sampling site was chosen to minimize fossil contamination (i.e., avoiding highways). The collected leaves were stored in perforated paper envelopes and refrigerated.

All sampling activities were carried out carefully to avoid $^{14}C$ contamination. Powder-free nitrile gloves were used during sampling. Plastic tubing (silicone, Teflon, and polypropylene) was rigorously cleaned with dilute HCl, and rinsed with ~10 L of distilled water. All other plastic ware was cleaned with soap and distilled water, leached with 10% (by volume) HCl in water, and then rinsed with distilled water. Glassware was cleaned in the same manner, followed by combustion at 450°C for > 4 hours. The R/V Blue Heron was free of $^{14}C$ contamination during a radiocarbon swipe test at the beginning of the project. Subsequent $^{14}C$ tracer work has been limited to a radiation van loaded and unloaded from the boat specifically for tracer work.

### 2.2.2 Concentration measurements

DOC concentrations were measured via high temperature catalytic combustion on a Shimadzu total organic carbon (TOC)-Vcsh analyzer, except for 2007 samples, which were analyzed by heated persulfate oxidation on an Oceanography International Corporation (OI) Analytical 1030 W TOC analyzer. Aliquots of the acidified DOC were bubbled with high grade $CO_2$-free air for 3.3 minutes to remove all inorganic carbon, 50 μL was then combusted at 680°C, and the evolved $CO_2$ was measured by a non-dispersive infrared (NDIR) detector. The TOC analyzer was calibrated using potassium hydrogen phthalate (KHP) and additional KHP standards were interspersed and analyzed along with the samples to assess instrumental performance. For each sample, three injections were performed. If the standard deviation was not ≤ 2.5%, two more injections were performed and the closest three of the five injections were averaged to yield sample concentration (Minor and Stephens 2008).

DIC samples were also analyzed on the Shimadzu TOC-Vcsh Analyzer. In this case, the analyzer was calibrated using primary standard grade sodium carbonate and ACS reagent grade sodium bicarbonate. The inorganic carbon in the sample was volatilized with 25% ACS grade $H_3PO_4$ (by weight) in a $CO_2$-free closed reaction vessel, and the $CO_2$ evolved was measured by the NDIR gas detector.

Suspended particulate organic matter (POM) samples were either freeze-dried or oven dried to constant weight at 60°C, and homogenized, fumigated with 12 mol $L^{-1}$ HCl (ACS Plus grade) overnight to remove carbonates, dried again, and cooled in a dessicator. They were then loaded into tin capsules and analyzed for particulate organic carbon (POC) and particulate organic nitrogen (PON) concentrations on a Costech ESC 4010 elemental analyzer.

### 2.2.3 Radiocarbon and stable carbon isotope measurements

All isotope measurements were performed at the National Ocean Sciences Accelerator Mass Spectrometry Facility (NOSAMS) at the Woods Hole Oceanographic Institution (WHOI), with the exception of the corn leaves sample, which was measured at the Keck Carbon Cycle AMS Facility (KCCAMS) at University of California, Irvine. Most water-column and porewater DOC samples were processed for radiocarbon measurement at NOSAMS by ultraviolet (UV) oxidation using a protocol based upon Beaupre et al. (2007) but oxidizing for three hours rather than four. DOC samples from September 2007 were freeze-dried before combustion to $CO_2$. In both approaches, the evolved $CO_2$ was trapped in a vacuum line, purified cryogenically and reduced to graphite with $H_2$ over Fe catalyst. A subsample of the purified $CO_2$ was collected for $\delta^{13}C$-DOC measurement.

Suspended particulate organic carbon and mesozooplankton biomass were oven-dried to constant weight at 60°C, fumigated with 12 mol $L^{-1}$ HCl (ACS Plus grade) for 24 hours to remove carbonates, re-dried, and combusted to $CO_2$ in a modified Carlo Erba NA1500 elemental analyzer. The evolved $CO_2$ was separated from the carrier gas, cryogenically trapped, and stored in a modular manifold, and then reduced to graphite. A subsample of the purified $CO_2$ was taken for $\delta^{13}C$-POC measurement.

DIC samples were directly hydrolyzed with $H_3PO_4$, and the resulting $CO_2$ was 'stripped' with nitrogen gas and trapped. The evolved $CO_2$ was cleaned cryogenically and reduced to graphite. A portion of the cleaned $CO_2$ was taken for $\delta^{13}C$-DIC measurement.

The radiocarbon content of atmospheric $CO_2$ was determined at KCCAMS using the protocol of Hsueh et al. (2007). The corn leaves sample was cleaned with ~5 L (in 5 separate rinses) of Milli-Q water to remove extraneous particles on the leaves, and dried to a constant weight for ~30 hours at 60°C. The dried sample was homogenized, combusted to $CO_2$ and graphitized as outlined in Santos et al. (2004). A subsample of the $CO_2$ from the corn leaves was taken for $\delta^{13}C$ analysis at KCCAMS using a gas bench coupled to a Finnigan Delta Plus Isotope Ratio Mass Spectrometer (IRMS).

In all cases, the graphite produced was compacted onto an aluminum cartridge target and analyzed by accelerator mass spectrometry (AMS) along with primary and secondary standards, and combustion and graphitization process blanks. Radiocarbon values are reported as $\Delta^{14}C$, the part per thousand deviation of the sample's $^{14}C$:$^{12}C$ ratio relative to a nineteenth century wood standard that has been corrected to the activity it would have had in 1950 and a $\delta^{13}C$ of -25‰. $\Delta^{14}C$ was corrected for fractionation using $\delta^{13}C$ of samples according to the convention of Stuiver and Polach (1977). Instrumental precision of $\Delta^{14}C$ analysis based on error of standards or multiple analyses on a target are explicitly given in Table 3, and ranged from 2‰ to 10‰. Total measurement uncertainties for $\Delta^{14}C$ analyses based on measurement of duplicate natural samples were 6‰ for $\Delta^{14}C$-DIC, and 16‰ for $\Delta^{14}C$-POC.

Combusting our large-volume blank 'POC' sample generated 94.75 μmol of C (or <0.95 μmol C $L^{-1}$), enough carbon for a radiocarbon analysis although this carbon would also include contributions from the MilliQ water as well as the filtration apparatus. The $\Delta^{14}C$ and $\delta^{13}C$ values of the large volume blank POC sample were -95 ± 3‰ and -27.3‰, respectively. For additional comparison, combusting a 'clean' GF/F filter generated 2.19 μmol of C. While this GF/F blank did not provide sufficient carbon for a radiocarbon analysis, it may be more representative of the amount of blank carbon in our samples, as it is not subject to the addition of carbon from 100 liters of MilliQ water. Using this GF/F filter blank and our lowest concentration POC sample, our processing techniques contribute 9% or less to the carbon identified in POC samples. Estimates of blank-carbon in our samples based upon the large-volume blank ranged from 6.0% to 37.7% of the carbon identified in our POC samples and the resulting radiocarbon correction would shift samples an average of 23‰ (with a standard deviation of 15‰) toward more $^{14}C$-enriched values. Basic depth and locational trends remained similar in the blank-adjusted and measured $\Delta^{14}C$ values but the extent of the depth variations is less strong in the blank-adjusted samples; note that blank-adjustment more strongly affects the radiocarbon values in the deep-water, low POC samples. Because we cannot be sure of the applicability of our blank carbon and radiocarbon values (as they contain carbon from the MilliQ water as well from the processing apparatus), we discuss measured POC $\Delta^{14}C$ values in the text that follows. The radiocarbon blank issue should be most severe with the POC samples (relative to DIC, DOC, and mesozooplankton samples) as they are by far the lowest in carbon concentration (see Table 2-1).

Stable carbon isotope ratios of samples (DOC, DIC, and POC) were measured at NOSAMS using an Optima stable isotope ratio mass spectrometer. Instrumental precision of $\delta^{13}C$ analysis was 0.15‰, and total measurement average deviation based on analyses of duplicate natural samples was 0.15‰ and 0.48‰, respectively, for $\delta^{13}C$-DIC and $\delta^{13}C$-POC (Table 2). All stable carbon isotope ratios ($^{13}C$:$^{12}C$) are reported as $\delta^{13}C$ which is the per mil difference relative to Pee Dee Belemnite carbonate. The δ notation is calculated as: $\delta Q = [R_{sample} / R_{standard} - 1] \times 1000$, where Q is $^{13}C$ and R is $^{13}C$:$^{12}C$. The $\delta^{13}C$-POC values are reported here as measured without correction using the large-volume blank data.

### 2.2.4 Proportion of pre-aged contents of POC ($POC_p$), and DOC ($DOC_p$)

$POC_p$ was estimated using a binary (pre-aged and modern) isotopic mixing model. The model is presented below, accounting for isotopic-mass balance (Eq. 2-1) and conservation of mass (Eq. 2-2)

$$\Delta^{14}C \bullet [POC] = \Delta^{14}C_m \bullet [POC_m] + \Delta^{14}C_p \bullet [POC_p] \quad (2\text{-}1)$$

$$[POC] = [POC_m] + [POC_p] \quad (2\text{-}2)$$

where the subscripts m and p refer to modern and pre-aged, respectively. We used the $\Delta^{14}C_{DIC}$ in the surface water as modern $\Delta^{14}C$ (and assigned this value to POM derived from recent photosynthesis, based on Druffel et al. 1996; Repeta and Aluwihare 2006). As a conservative estimate, the pre-aged end-member was assigned a $\Delta^{14}C$ value of -24‰ (173 yr BP), equivalent to the hydrologic residence time (the average number of years that water stays in the lake based on volume divided by average net outflow rates) of the lake. Using our measured suspended $\Delta^{14}C_{POC}$ values and POC concentrations, we calculated $POC_p$ within bulk suspended POM at each depth. For comparison (as a sensitivity analysis), we also calculated $POC_p$ using our large-volume-blank-adjusted values for $\Delta^{14}C_{POC}$. $DOC_p$ was calculated similarly, using measured [DOC] and $\Delta^{14}C$ and the same estimates for $\Delta^{14}C_m$ and $\Delta^{14}C_p$ as in the $POC_p$ calculation.

### 2.2.5 Proportion of terrigenous content of POC ($POC_T$)

$POC_T$ was estimated using a binary (terrigenous and autochthonous) mixing model as follows:

$$C{:}N_{POM} = f \bullet C{:}N_{Terr} + (1 - f) \bullet C{:}N_{Auto} \qquad (2\text{-}3)$$

where $f$ is the fraction of terrigenous organic carbon in the suspended POC, $(1 - f)$ is the fraction of autochthonous organic carbon in the suspended POC, and the subscripts ‘Terr’ and ‘Auto’ refer to terrigenous and autochthonous, respectively. We used a molar C:N ratio of 6.6 (based on Redfield ratio, Redfield et al. 1963) as representative of autochthonous OM sources. While we were unable to find values for the C:N ratio within Lake Superior phytoplankton, the range of ratios found within particulate organic matter from Lake Superior (4.6 to 14.2, Sterner 2010) indicates that the Redfield ratio is a good first approximation. We applied a molar C:N ratio of 21 for the terrigenous end-member.

## 2.3 Results

### 2.3.1 Concentration of carbon pools

The concentrations of DOC, POC, and PON showed temporal (seasonal) and spatial (lateral and depth profile) variations (Table 2-1). At the OL site, two distinct seasonal depth profiles of DOC, POC, and PON are distinguishable: one during thermal stratification, and the other during isothermal or mixed conditions.

During isothermal conditions, the nearshore site (BR) exhibited slightly higher DOC concentrations than the surface waters at the offshore site (OL) (Table 2-1). Similarly, the POC and PON concentrations were higher at BR than the surface waters of OL (Table 2-1). At site OL, the concentrations in the isothermal condition were low and invariant across depth, with water column surface-to-deep-water averages (and standard deviations) of 81.1 (± 0.9) $\mu mol\ L^{-1}$ of DOC, 6.9 (± 0.7) $\mu mol\ L^{-1}$ of POC, and 0.8 (± 0.1) $\mu mol\ L^{-1}$ of PON in May 2008, and 93.6 (± 0.5) $\mu mol\ L^{-1}$ of

DOC, 4.8 (± 0.1) µmol $L^{-1}$ of POC, and 0.6 (± 0.1) µmol $L^{-1}$ of PON in June 2009 (Table 2-1).

During stratification, DOC concentrations at BR (107 µmol $L^{-1}$, and 98 µmol $L^{-1}$, respectively, in September 2008, and August 2009) were similar to the values at surface OL (102 µmol $L^{-1}$ in September 2008 and 104 µmol $L^{-1}$ in August 2009) (Table 2-1). POC values at BR (11.5 µmol $L^{-1}$, and 13.4 µmol $L^{-1}$) during the stratified period were similar to those seen in surface (15.9 µmol $L^{-1}$, and 12.0 µmol $L^{-1}$), and chlorophyll maximum waters (11.9 µmol $L^{-1}$, and 9.6 µmol $L^{-1}$) at the OL site (Table 2-1). For all samplings, surface concentrations of DOC, POC and PON at site OL were higher during stratification than during isothermal periods (Table 2-1).

DIC concentrations in the water column at OL did not vary much with depth during isothermal condition. The DIC values ranged from 809 µmol $L^{-1}$ to 819 µmol $L^{-1}$ in May 2008, and 822 µmol $L^{-1}$ to 830 µmol $L^{-1}$ in June 2009 (Table 2-1). In the stratified months, DIC concentration increased with depth from 802 µmol $L^{-1}$ at the surface to 830 µmol $L^{-1}$ in the deep waters in September 2008, and from 807.5 µmol $L^{-1}$ to 831 µmol $L^{-1}$ in August 2009 (Table 2-1).

The molar C:N ratios of suspended POM ($C{:}N_{POM}$) in the lake from 2007 to 2009 did not exhibit consistent variation with depth or season (Table 2-1). Overall, the $C{:}N_{POM}$ values in the lake ranged from 7.3 to 9.3, and 7.1 to 9.5, respectively, at the OL and BR sites.

## 2.3.2 Stable isotopic composition

The $\delta^{13}C$ of DIC ($\delta^{13}C_{DIC}$) values were $^{13}C$-depleted progressively with depth at site OL during the stratified period. The $\delta^{13}C_{DIC}$ values decreased from 1.2‰, 1.5‰, and 1.3‰ (average of 1.3 ± 0.1‰) in the surface waters to 0.4‰, 0.2‰, and 0.2‰ (average of 0.2 ± 0.1‰) in the deep waters, respectively, in September 2007, September 2008, and August 2009 (Table 2-2; Fig. 2-2). In the isothermal water column in spring, $\delta^{13}C_{DIC}$ values at site OL were similar in both surface waters (0.2‰ in May 2008, and 0.4‰ in June 2009) and deep waters (0.5‰ in May 2008, and 0.3‰ in June 2009), thus giving water column averages of 0.4 ± 0.1‰ in May 2008, and 0.3 ± 0.1‰ in June 2009. The $\delta^{13}C_{DIC}$ at BR was more $^{13}C$-enriched in late summer (1.2‰, 1.2‰, and 0.9‰, respectively, in September 2007, September 2008, and August 2009) than in spring (0.5‰, and 0.3‰, respectively, in May 2008, and June 2009) (Table 2-2).

The $\delta^{13}C_{DIC}$ at BR was similar to that of surface OL samples during the stratified seasons and to all OL samples during the isothermal samplings.

The distribution of $\delta^{13}C_{POC}$ at site OL changed markedly with depth during the stratified seasons. Stratified-season $\delta^{13}C_{POC}$ values at OL ranged from -27.8‰ to -26.7‰ in the surface waters, -29.7‰ to -29.4‰ in the DCM, and -29.8‰ to -29.5‰ in the deep waters (Table 2-2). In September 2008 and August 2009, the most $^{13}C$-depleted $\delta^{13}C_{POC}$ values were observed in the DCM (Table 2-2). During spring isothermal condition, the $\delta^{13}C_{POC}$ values were nearly invariant with depth, averaging -30.5 ± 0.4‰ in May 2008, and -29.5 ± 0.3‰ in June 2009 (Table 2-2). POC from the surface waters in the isothermal period exhibit more $^{13}C$-depleted values than the corresponding stratified-period surface samples. Our data show a positive correlation between $\delta^{13}C_{POC}$ and $\delta^{13}C_{DIC}$ ($R^2 = 0.641$, $p < 0.001$, $n = 21$), with the relationship strongly driven by stratified samples, especially those from surface water (at both BR and OL sites) and the DCM (Fig. 2-3).

The $\delta^{13}C_{DOC}$ varied from -27.5‰ to -25.7‰ in the well-mixed lake at site OL in May 2008 and June 2009 (Table 2-2). In the stratified lake, the September 2007 DOC samples, which were freeze-dried, were more $^{13}C$-enriched (surface to deep water ranged from -25.3‰ to -19.1‰) than the samples from September 2008 (ranged from -25.9‰ to -25.6‰), and August 2009 (ranged from -26.1 to -25.8‰). Our data show no distinct $\delta^{13}C_{DOC}$ depth profiles during the stratified period, in contrast to $\delta^{13}C_{POC}$ (Table 2-2). The $\delta^{13}C_{DOC}$ values at the BR site (-28.9‰ in September 2008, and -26.5‰ in August 2009), are more $^{13}C$-depleted relative to the corresponding values in the surface waters at OL (-25.6‰, and -26.1‰, Table 2-2).

### 2.3.3 The $\Delta^{14}C$ signatures of carbon pools

The $\Delta^{14}C$ value of atmospheric $CO_2$ ($\Delta^{14}C_{Atm\ CO2}$) in the Lake Superior watershed as integrated in the leaves of an annual plant (corn) was 38 ± 2‰ in September 2009. Hsueh et al. (2007) reported a $\Delta^{14}C$ value of 65 ± 2‰ for corn leaves collected at the western basin of the lake in summer 2004, implying that the mean annual decline rate of $\Delta^{14}C_{Atm\ CO2}$ was 4.5‰ to 6.0‰ $yr^{-1}$ over the past 5 years. The $\Delta^{14}C_{DIC}$ values in the surface waters at site OL during the stratified period were 83‰, 69‰, and 61‰, respectively, in 2007, 2008, and 2009 (Table 2-3, Fig. 2-4) revealing a mean $\Delta^{14}C$ decline rate of ~11‰ $yr^{-1}$. The $\Delta^{14}C_{DIC}$ values in the deep waters at OL showed a similar decline from 74‰ in 2007 to 52‰ in 2009. There were no significant variations in $\Delta^{14}C_{DIC}$ as a function of depth within the current precision of AMS (2-7‰, Turnbull et al. 2006; McNichol and Aluwihare 2007) implying thorough mixing between the surface and deep water DIC.

The $\Delta^{14}C$ values of POC at site OL varied as a function of depth (Fig. 2-2, Table 2-3). POC in the surface waters during stratification (61‰, 48‰, and 33‰, respectively, in September 2007, September 2008, and August 2009) was more $^{14}C$-enriched than corresponding deep water POC (12‰, 10‰, and -17‰, respectively, in September 2007, September 2008, and August 2009), and closer to co-occurring $\Delta^{14}C_{DIC}$ levels (Table 2-3). The $\Delta^{14}C$ values of deep water POC were $^{14}C$-depleted to a greater extent than the co-occurring DIC in the stratified lake. The decline rate of surface-water $\Delta^{14}C_{POC}$ in the stratified lake from 2007 to 2009 (~14‰ $yr^{-1}$) was similar to that of $\Delta^{14}C_{DIC}$. Correlation analysis indicates a positive linear relationship between the $\Delta^{14}C$ in POC and DIC at site OL, with 41% (based on coefficient of determination, $R^2$ of 0.41) of $\Delta^{14}C_{POC}$ variation predicted by $\Delta^{14}C_{DIC}$.

Measured $\Delta^{14}C_{POC}$ values were generally more $^{14}C$-depleted than $\Delta^{14}C_{DOC}$. Exceptions include one deep water sample in September 2007 in which the DOC was more $^{14}C$-depleted (-77‰) and two samples from the May 2008 isothermal sampling, in which the DOC and POC were not significantly different in $\Delta^{14}C$. (Large-volume-blank-adjusted $\Delta^{14}C_{POC}$ values were closer to those of $\Delta^{14}C_{DOC}$, but the blank-adjusted $\Delta^{14}C_{POC}$ remains $^{14}C$-depleted relative to the corresponding measured $\Delta^{14}C_{DOC}$ values in many of the samples). The DOC was more $^{14}C$-enriched (51‰ to 73‰) in the surface waters and $^{14}C$-depleted in the deep waters (-77‰ to 14‰) during the stratified periods in 2007 to 2009 (Table 2-3).

Correlation between $\Delta^{14}C_{DOC}$ and $\Delta^{14}C_{DIC}$ indicated a positive relationship with ~52% ($R^2 = 0.52$) of the $\Delta^{14}C$ variation in DOC accounted for by DIC when the anomalous deep water $\Delta^{14}C_{DOC}$ value of -77‰ was excluded. The DOC in the surface waters at OL during stratification in September 2008 and August 2009 was more $^{14}C$-depleted compared to DOC from the nearshore BR site (Table 2-3).

### 2.3.4 Sedimentary porewater DOC

At site OL, concentrations of DOC were somewhat higher in the surficial sediments (0-2 cm) in June 2009 (108 μmol $L^{-1}$), and much higher in August 2009 (395 μmol $L^{-1}$) than in the immediately overlying water in direct contact with the sediments (102 μmol $L^{-1}$ in June 2009 and 118 μmol $L^{-1}$ in August 2009) (Table 2-4). The porewater DOC from surficial sediments (0-2 cm) was considerably depleted in $^{14}C$ relative to the water column in the lake. The $\Delta^{14}C$ and $\delta^{13}C$ values of the porewater DOC (0-2 cm) were, respectively, -302‰ and -35.6‰ in June 2009, and -158‰ and -33.3‰ in August 2009 (Table 2-4).

### 2.3.5 Amity Creek

We determined the concentrations and isotopic compositions of Amity Creek carbon coming into the lake in order to assess its varying nature and thus potential allochthonous inputs from North Shore streams into western Lake Superior. Amity Creek DOC concentrations were 489 μmol $L^{-1}$ during baseflow in September 2008 and as high as 885 μmol $L^{-1}$ at the same sampling location during a storm event in June 2008 (Table 2-5). Baseflow DIC concentration at the creek was 2421 μmol $L^{-1}$, and DIC was the largest carbon pool at this sampling time.

The $\delta^{13}C$ composition of DOC in the creek (-28.6‰ to -28.1‰) did not vary much as a function of flow rate (Table 2-5). In contrast, the $\Delta^{14}C$ of DOC was $^{14}C$-enriched with flow rate, increasing from -13‰ during baseflow in September 2008 to 62‰ during stormflow in June 2008. The $\delta^{13}C_{DIC}$ and $\Delta^{14}C_{DIC}$ values were -10.9‰ and 21‰, respectively, during baseflow in September 2008 (Table 2-5). The DIC was $^{14}C$-depleted by ~18‰ relative to the $\Delta^{14}C$ value of atmospheric $CO_2$. The POC pool was more $^{14}C$-depleted than creek DOC during the June 2008 storm event.

### 2.3.6 Mesozooplankton

The $\delta^{13}C$ values of mesozooplankton biomass ($\delta^{13}C_{Zoops}$) were -29.7‰ and -31.0‰ in June and August 2009, respectively (Table 2-6). The $\Delta^{14}C_{Zoops}$ values (57‰ and 62‰, Table 2-6) were nearly the same as $\Delta^{14}C_{DIC}$ values (56‰ and 61‰) in June and August 2009, but more $^{14}C$-enriched than co-occurring POC.

### 2.3.7 Pre-aged, and terrigenous carbon compositions

Based upon our mixing model and measured $\Delta^{14}C_{POC}$ values, the proportion of pre-aged materials in the surface water POC samples during stratification were 20%, 22%, and 32%, respectively, in September 2007, September 2008, and August 2009. These values increased, respectively, to 66%, 63%, and 92% in the deep water POC (Table 2-7). Using large-volume-blank-adjusted $\Delta^{14}C_{POC}$ values shifts the pre-aged proportion in both surface-water and deep-water POC during stratification and increases the scatter, with surface-water values of 3%, 13%, and 22% and deep-water values of 6%, 33%, and 65%. The proportions of pre-aged materials at BR, based upon measured values, were 17% in September 2008, and 63% in August 2009 (Table 2-7), and, based upon large-volume-blank corrected values, were 3% and 35%, respectively.

For the DOC, during stratification the pre-aged content increased from 9% (September 2007) and 11% (August 2009) in the surface water to values ranging from 55% (August 2009) to 74% (September 2008) in the deep water (Table 2-7). The DOC in the isothermal water column showed a similar proportion of pre-aged carbon throughout the water column at site OL (Table 2-7) consistent with what would be expected from water column mixing. The pre-aged proportion of POC showed a less strong relationship with depth during isothermal conditions, but was higher in surface waters in comparison to the stratified periods.

The contribution of terrestrial carbon to the POC pool at site BR from four samplings from September 2008 to August 2009 ranged from 0.9 to 1.6 μmol C $L^{-1}$ (Table 2-7), and was consistently higher relative to the samples from OL as would be expected due to proximity to shore (Table 2-7). This, however, did not necessarily translate into a higher proportion of terrigenous carbon due to correspondingly high aquatic primary productivity at site BR. The terrestrial carbon concentrations at OL during stratification decreased with depth in the water column, though the percentage of POC considered terrestrial increased from the DCM to the deep water. The averages of concentrations of terrestrial carbon from the three stratified samplings in September 2007, September 2008, and August 2009 were: 2.0 ± 0.9 μmol C $L^{-1}$ (16 ± 3%) in the surface water POC, 0.9 ± 0.2 μmol C $L^{-1}$ (9 ± 2%) in the DCM, and 0.5 ± 0.2 μmol C $L^{-1}$ (16 ± 1%) in the deep water POC (Table 2-7). Our calculated water column (surface-to-deep water) average values for terrigenous carbon subsidy to the open lake POC was 13 ± 4% during late summer stratification for the three stratified periods (August 2007, September 2008, and August 2009) and 9 ± 3% during the two spring mixing periods (May 2008 and June 2009). There was a positive relationship between estimated terrestrial carbon input and $\Delta^{14}C_{POC}$ at the open-lake site ($R^2 = 0.31$, $p < 0.01$, $n = 15$) (Fig. 2-5C).

# 2.4 Discussion

## 2.4.1 Concentration trends of DOC, POC, PON, and DIC

The somewhat higher DOC and POC concentrations at the nearshore site (BR) relative to the offshore site (OL) in May or June is due most likely to a combination of allochthonous OM inputs, the effects of terrestrially-derived nutrient-loading on autochthonous primary productivity (Urban et al. 2005), and the effects of nearshore increases in water temperature on productivity. The open-lake DOC concentrations agree well with previously reported ranges (Table 2-8). During thermal stratification, the increase in DOC concentration in the surface waters at OL relative to that in the isothermal condition is possibly due to increased exudation from phytoplankton cells and release from sloppy feeding by zooplankton. The considerably lower concentrations of DOC, POC and PON in the deep cold waters during stratification conceivably reflect a background level of soluble and particulate OM that persists in the water column across seasons.

DOC concentration in the lake has been fluctuating seasonally and within season (monthly) in the past four decades (Table 2-8). In addition to temporal changes, DOC concentration is likely to differ spatially within the lake due to water mass movement and changes in biogeochemical processes. The lake has not yet been sampled at fine enough resolution to determine spatial patchiness in DOC concentrations resulting from currents, eddies, and other mesoscale phenomena. In addition, most carbon concentration measurements have been performed on samples obtained during fair weather conditions conducive for sampling from a research vessel. Over the past few decades, lake DOC shows concentrations oscillating within a range of 80-210 μmol $L^{-1}$ (Table 2-8). An interesting observation in our data set was a sudden significant decline (~36 %) in DOC concentration between September 2007 and May 2008. This is consistent with values reported by Sterner (2010), indicating a similar decline of ~32 % of DOC values between July 2007 and July 2008. Sterner (2010) also reported ~40 % decline in DOC concentration over just a period of 4 months (April 2008 to July 2008) (Table 2-8). These rapid declines of the largest organic carbon pool in the lake are very intriguing, and are likely driven by within-lake biogeochemical processes rather than lake water-level changes. They could be a key reason for the imbalanced carbon budgets reported by Cotner et al. (2004) and Urban et al. (2005), since the DOC pool is likely to be poorly constrained by snapshot data.

The POC concentrations are consistent with values (≤ 18 μmol $L^{-1}$) in Lake Superior over the past 4 decades and appear to show seasonal variation, with low values in spring (Table 2-8). The PON concentrations are significantly lower than DOC and POC, and consistent with values reported by Sterner (2010), but slightly higher than the values of 0.1 to 0.4 μmol $L^{-1}$ reported by Ostrom et al. (1998). The increased surface concentration of POC and PON at OL is potentially a result of increased algal, bacterial and zooplankton production (Velinsky and Fogel 1999). Historically, the POC concentrations in the lake are quite variable, but may have declined slightly in the past few decades (Table 2-8).

Our DIC concentrations are within the range (733-965 μmol $L^{-1}$, Table 2-8) seen in the lake over the past 4 decades. Reported DIC values in the past 4 decades exhibited some seasonal variation, but have been generally steady (Table 2-8). In our study (Table 2-1), DIC concentrations were higher in the surface waters during isothermal conditions, likely as a result of 1) lower primary productivity, 2) increased surface water exchange with the atmosphere (Cole et al. 1994), and 3) higher $CO_2$ solubility at low temperatures. During stratified conditions (Table 2-1), we see an increase in DIC with depth, probably due to utilization of surface DIC for photosynthesis, mineralization of sinking particulate matter, and the increased solubility of $CO_2$ in the cooler deep waters.

The C:N values of suspended POM (Table 2-1) are consistent with previously reported values in the lake. Halfon (1984) reported an average C:N value of 8.4 for suspended POM at various depths (≤ 100 m); Guildford and Hecky (2000) reported a C:N range of 7.2-12.7 (mean of 10.3) for suspended POM; and Urban et al. (2004a,b) reported a value of 8.5 ± 0.3 for suspended POM in the upper 25 m off the Keweenaw Peninsula region of Lake Superior. The C:N values of suspended POM from 2-80 m in the water column (C:N values of 7.8-14.3) reported by Sterner (2010) for 2007-2008 are in agreement with our values, but the values reported for 2006 (4.6-7.9; Sterner 2010) are considerably lower than our values. The lower C:N values in 2006 seen by Sterner (2010) could be due to varying internal processes within the lake, or considerably lower terrestrial influence on POM during that year.

Our $C:N_{POM}$ values do not reflect purely terrigenous plants (C:N > 20) or solely phytoplankton (C:N ~6.6; Redfield et al. 1963) sources of OM. They are consistent with a composite of the two, with the algal proportion being predominant (Meyers 2003), and are also at the lower range of C:N found in cool biome soils (Aitkenhead and McDowell 2000). In comparison with open oceanic systems, our C:N values are higher than values reported for suspended POM in equatorial Pacific Ocean (5.7; Pena et al. 1991) and the equatorial Atlantic Ocean (6.1; Herbland and Le Bouteiller 1981), probably as a result of greater riverine inputs (and their concomitant inputs of terrestrially-derived material) relative to water volume.

### 2.4.2 Stable isotopic composition of inorganic and organic carbon pools

Our $\delta^{13}C_{DIC}$ values are in general agreement with other reported values (-4‰ to 0‰) in the lake (Keough et al. 1996; Hoffman et al. 2010). The $\delta^{13}C_{DIC}$ values at site OL during stratified conditions are consistent with depletion of $^{12}C$ due to photosynthetic fractionation in surface waters, and subsequent respiration of sinking OM which releases $^{12}C$-enriched $CO_2$ into deep waters (Velinsky and Fogel 1999). Similarity in surface and deep $\delta^{13}C_{DIC}$ values at site OL during isothermal conditions in spring is due to homogenization of the water column from vertical mixing.

The $\delta^{13}C_{POC}$ values in the lake during the stratified season are consistent with values previously observed in far western coastal Lake Superior (Hoffman et al. 2010). The $\delta^{13}C_{POC}$ values in the surface water (-27.8‰ to -26.7‰) during stratification are similar to the $\delta^{13}C$ of primary producers (-27.9‰ to -26.7‰ for diatoms, and -26.7‰ to -26.3‰ for chrysophytes; Keough et al. 1996, 1998) in the lake. Sierszen et al. (2006) also reported that small-sized plankton (110-253 μm) in the lake had $\delta^{13}C$ values of -28.7‰ to -26.9‰, whereas $\delta^{13}C$ of large-sized plankton (> 183 μm) was -28.2 ± 1.2‰. These values are similar to the $\delta^{13}C_{POC}$ values in the surface water of the lake during the stratified season. In the stratified lake in September 2008 and August 2009, the lowest $\delta^{13}C_{POC}$ value was observed in the DCM, similar to observations in the NE Pacific Ocean (Bianchi et al. 1998; Druffel et al. 1998).

The $^{13}C$-depletion of POC with depth observed in Lake Superior is consistent with observations in the Middle Atlantic Bight shelf and slope in July - August, 1996 (Bauer et al. 2002), but contrasts with $^{13}C$-enrichment of POC with depth observed in North East (NE) Pacific in September 1994 and June 1995 (Druffel et al. 1998). These shifts in $\delta^{13}C$ between depths in Lake Superior are most likely due to slower algal growth rates in the colder sub-surface depths (Karlsson et al. 2003; Marty and Planas 2008). Other potential causes could be shifts in photoautotrophic community composition (Falkowski 1991; Velinsky and Fogel 1999), incorporation of dissolved $CO_2$ vs. bicarbonate by different algal species (Laws et al. 1997), variation in temperature, light intensity and nutrients (Freeman and Hayes 1992; Burkhardt et al. 1999; Hayes 2001), variation in cell geometry (Popp et al. 1998), or a shift in the sources of particles to the deep vs. surface waters.

The $^{13}$C-depletion of POC values in the isothermal periods in May 2008 and June 2009 relative to the following stratified-period samples (September 2008 and August 2009) is consistent with low temperature (~4°C) and reduced phytoplankton growth rates in the colder spring samplings. The trends in Fig. 2-2E are consistent with autochthonous processes as a primary driver of POC stable isotope composition. The portion of the $\delta^{13}C_{POC}$ signature that is not correlated to $\delta^{13}C_{DIC}$ (Fig. 2-3) suggests that additional processes may be affecting the $\delta^{13}C_{POC}$ signal in the surface and in the deep waters; one process likely to be significant is sedimentary resuspension, as discussed below.

The $\delta^{13}C_{DOC}$ values are consistent with $\delta^{13}C$ values of phytoplankton in the lake (Keough et al. 1996, 1998; Sierszen et al. 2006). The variation in $\delta^{13}C_{DOC}$ in the stratified lake is limited to the freeze-dried samples from September 2007. These were more $^{13}$C-enriched than the DOC samples from September 2008 and August 2009, which were not freeze-dried and combusted, but rather UV-oxidized in their liquid state. It is not clear whether this $^{13}$C-enrichment is due to unique sources, in-lake variability, or the freeze-drying process. The consistent $^{13}$C-enrichment of DOC (both freeze-dried samples and wet-oxidized ones) relative to POC may indicate that POC and DOC are differently sourced (Raymond and Hopkinson 2003) and/or that the DOC pool was relatively more subjected to photodegradation (Opsahl and Zepp 2001; Vahatalo and Wetzel 2008), or a combination of these factors.

### 2.4.3 The $\Delta^{14}C$ within atmospheric $CO_2$, water column DIC, and water-column POC

Our data show a consistent $^{14}$C-depletion of atmospheric $CO_2$ relative to surface DIC from 2007 to 2009; this is due to positioning on the tailing portion of the 'bomb spike' (cf, Fig. 5 in McNichol and Aluwihare 2007). The absence of a reservoir effect in the lake implies that the hardwater effect is negligible. In other words, there is little to no $^{14}$C-depletion of the DIC pool relative to atmospheric $CO_2$ due to addition of radioactively inert

($^{14}C$-dead or old) carbon from limestone and/or dolomite in the sediments or rocks in the lake basin. This absence of a reservoir effect is consistent with the lake's large surface area, the absence of significant limestone and dolomite deposits in its drainage basin, and the biannual vertical mixing of the lake water column.

The offset between radiocarbon values in DIC and atmospheric $CO_2$ indicates that the DIC pool is not in equilibrium with atmospheric $^{14}CO_2$. Potential causes of this include: 1) Reduction in the rate of exchange between water column DIC and atmospheric $CO_2$ during periods (April and November; Urban et al. 2005) when the lake is supersaturated with $pCO_2$ and effluxes $CO_2$ to the atmosphere instead of taking in atmospheric $^{14}CO_2$ (Abbot and Stafford 1996), 2) Reduction in exchange between lake-DIC and atmospheric $^{14}CO_2$ during winter ice cover (December to May), 3) Relatively long residence time of lake-DIC. We determined the residence time to be ~3 years (calculated using the DIC pool, ~125 Tg C, divided by lake-atmospheric $CO_2$ exchange flux, $F_{CO2}$ of 50 Tg C $y^{-1}$; Alin and Johnson 2007). Our $\Delta^{14}C_{DIC}$ values are consistent with this concentration-and-flux-based estimate showing $^{14}C$-enrichment by ~20‰ relative to the $\Delta^{14}C_{Atm\ CO2}$ in 2009.

The similarity in decline rate of $\Delta^{14}C_{POC}$, and $\Delta^{14}C_{DIC}$ values in the surface waters during stratified conditions, in conjunction with our C:N data, and $\delta^{13}C$ trends for both POC and DIC, implies that much of the POC in the stratified surface waters is derived from recent photosynthesis in the lake. However, contrary to the surface waters, $\Delta^{14}C$ values of deep water POC were $^{14}C$-depleted to a greater extent than the co-occurring DIC in the lake during stratification, consistent with inputs of older POC from sediment resuspension and with selective diagenetic remineralization or solubilization of younger POC. Ageing during vertical transport alone (i.e., as a function of the time of sinking) cannot account for the $\Delta^{14}C$ depth gradient.

The positive correlation observed between $\Delta^{14}C$ in POC and DIC could most likely be due to aquatic photosynthetic depletion of DIC, and the subsequent homogenization of fixed DIC throughout the water column via particle-sinking and the semi-annual water-column overturn. The remainder of the POC $\Delta^{14}C$ signal that is not explained by DIC is most likely a function of inputs from the surrounding watershed, sediment resuspension events, and the presence of trace black carbon from forest fires and fossil fuel combustion.

The general $^{14}C$-depletion in POC relative to DOC, whose $\Delta^{14}C$ values are closer to those of the DIC pool, indicates a different combination of sources (Raymond and Hopkinson 2003) and a smaller proportional contribution of fresh primary production (both allochthonous and autochthonous) to the POC relative to the DOC in the lake. Our observations contrast with observations in most open oceans where the DOC pool is consistently much older than the suspended POC (Druffel et al. 1992; Druffel and Bauer 2000) but agree with those from smaller lakes (S. L. McCallister unpubl.). This contrast is likely due to the extent of fluvial influence (in which land-derived DOM from the past 50 years of terrestrial primary production will be labeled with bomb radiocarbon) and the physical homogenization of the water column in the spring and early winter. Enhanced particle resuspension in the lake, with its shorter water column and vigorous storm events, could also contribute to older POC values in the subsurface water column. The depth gradient in $\Delta^{14}C_{DOC}$ during the stratified period as a result of $^{14}C$-enrichment in the surface waters, and $^{14}C$-depletion in deep waters, is possibly due to input of $^{14}C$-enriched DOC from photosynthesis in the surface water, and also an influx of $^{14}C$-depleted DOC from sedimentary porewater into the overlying deep water.

### 2.4.4 Comparison of $\Delta^{14}C$ within water column DOC and POC and porewater DOC

The DOC in the surface waters at OL during stratification in September 2008 and August 2009 was more $^{14}C$-depleted compared to DOC from the BR site. This difference is most likely due to larger modern DOC input at site BR resulting from enhanced photosynthesis driven by greater nutrient concentrations and higher temperatures.

$\Delta^{14}C_{DOC}$ values were relatively less variable (39‰ to 59‰) during vertical mixing in May 2008 and June 2009 reflecting the $\Delta^{14}C$ signal of a composite DOC pool derived from mixing of $^{14}C$-enriched (51‰ to 75‰) DOC in the surface water, and $^{14}C$-depleted (-77‰ to 14‰) DOC in the deep water. It is interesting to note that mean $\Delta^{14}C$ within the composite DOC pool (46‰ ± 8‰) during spring mixing was $^{14}C$-depleted by ~20‰ relative to mean $\Delta^{14}C_{DOC}$ in the surface water during stratification, but $^{14}C$-enriched by ~77‰ relative to mean $\Delta^{14}C_{DOC}$ in the deep waters during stratification. This suggests a stronger role for semi-reactive DOC, and a tighter coupling between DOC inputs and removal, in Lake Superior compared to most open ocean systems where the pre-aged non-reactive

DOC dominates the bulk $\Delta^{14}C_{DOC}$ signal (Druffel et al. 1992; Druffel and Bauer 2000). This is not surprising considering the highly varying water residence times (~173 years for Lake Superior vs. ~1000-1500 years for the deep ocean [Urban et al, 2005; McNichol and Aluwihare 2007]) and ventilation ages (months in Lake Superior vs. ~1000 years in the ocean) in the two systems.

Our Lake Superior porewater DOC concentrations (93-395 μmol C $L^{-1}$) are lower than reported values for surface sediments in Lake Michigan (900 μmol $L^{-1}$ ; O'Loughlin and Chin 2004), and Lake Ontario (2600 μmol $L^{-1}$; Roy and Knowles 1994). This is not surprising since Lake Michigan and Lake Ontario are both more terrestrially influenced and autochthonously productive, and have higher sedimentation rates than Lake Superior (Cotner et al. 2004). The difference in concentration between overlying water and porewater (0-2 cm) at site OL in Lake Superior indicates diffusion of porewater DOC into the overlying water column (Alperin et al. 1994; Burdige 2002).

The $^{14}C$-and $^{13}C$-enrichment of porewater DOC in August relative to June is probably due to enhanced fluxes of $^{14}C$- and $^{13}C$-enriched POM sinking to the sediment-water interface from the surface water in August, and a lower concentration-gradient induced flux from the deeper porewaters. The $^{14}C$-depletion in the porewater relative to the overlying water column is likely due to old carbon and/or preferentially preserved refractory components (e.g., bound- and free lipids, and lignocelluloses; Burdige 2007) diffusing from the deeper sediments into the surficial layers.

## 2.4.5 The $\Delta^{14}C$ and $\delta^{13}C$ distributions in carbon from Amity Creek

Our DOC concentrations are within the range of reported values (230-3000 μmol $L^{-1}$) for streams and rivers entering Lake Superior (Maier and Swain 1978; Urban et al. 2005). DOC constitutes approximately 17% of the total carbon exported by Amity Creek during baseflow conditions (compared to 83% from DIC and only trace POC).

Observed $\delta^{13}C_{DOC}$ values in Amity creek are consistent with OM from terrestrial and riverine C3 plants and do not vary as a function of flow or discharge rate. Our $\Delta^{14}C$ values show that pre-aged DOC (45 ± 35 years BP) is drained from the watershed during low flow conditions. This pre-aged DOC could be from ground water DOM, OM derived from fixation of groundwater DIC, incorporation of OM fixed prior to the 1950s (perhaps from deeper soil layers), or addition of fossil-derived carbon. The

DOC during storm flows was $^{14}$C-enriched, indicating that most of this DOC originated from recent photosynthesis and most likely came from the uppermost soil layers. The $^{14}$C-depletion of POC relative to co-occurring DOC during peak flows is consistent with observations in northeast US rivers [e.g., Hudson, Delaware and Parker rivers] (Raymond et al. 2004), and in temperate estuaries (Raymond and Bauer 2001; Raymond and Hopkinson 2003). The observed POC-DOC $\Delta^{14}$C offset suggests that a larger proportion of the POC is derived from 'older' particulates stored in the watershed soil and rock or within the stream bed for longer periods (Raymond and Hopkinson 2003), from pre-bomb woody material, or from some addition of fossil-derived particulate black carbon incorporated into the soils (Penner et al. 1993; Massiello and Druffel 1998).

The $\delta^{13}C_{DIC}$ value observed in Amity Creek is consistent with reported values of $\delta^{13}C_{DIC}$ in Bluff Creek (-11.9‰; Keough et al. 1998) and Bear Creek (-12.9‰; Keough et al. 1998), which are wetland tributaries in the western Lake Superior watershed. The important sources and processes controlling the isotopic signal of DIC in the stream are exchange with atmospheric $CO_2$ ($\delta^{13}C$ = -7.8‰; Freeman and Hayes 1992), $CO_2$ from soil respiration ($\delta^{13}C$ = -29.0‰ to -25.0‰; Raymond and Hopkinson 2003), groundwater input, proportion of $CO_{2(aq)}$ vis-à-vis $HCO_3^-$ ($\delta^{13}C$ = 1.0‰; Laws et al. 1997) species, and DIC influx from soil ($\delta^{13}C$ ~ -12.0‰), especially during spring snowmelt or ice-out (Striegel et al. 2001). For our stream site, DIC in equilibrium with atmospheric $CO_2$ would be relatively $^{13}$C-enriched (– 1.0‰ to 1.0‰) whereas $CO_2$ produced from soil respiration would be $^{13}$C-depleted by 14 to 18‰ relative to our observed value of ~ -11‰. A binary mixing model using the $\delta^{13}C$ of atmospherically-fixed aqueous $CO_2$ and soil respiration $CO_2$ as end-members indicates that OM decomposition could contribute from 38% to 43% of the total stream DIC.

The $^{14}$C-depletion of DIC relative to the $\Delta^{14}$C value of atmospheric $CO_2$ implies some $CO_2$ input from 'old' (i.e., pre-bomb) inorganic or organic carbon in the watershed, or groundwater input. As there is no limestone in this region, we applied a mixing model using the $\Delta^{14}$C of atmospheric $CO_2$ and soil respiration $CO_2$ as end-members, assuming the old OM respired had a $\Delta^{14}$C value that is same as the pre-aged baseflow DOC ($\Delta^{14}$C = -13‰). This calculation indicates that approximately 34% of the DIC originated from soil respiration. This estimate, much less well constrained due to variable radiocarbon ages within soil OM fractions and soil horizons (Crow et al. 2009; Koarashi et al. 2009), and the estimate from

$\delta^{13}C$ values, both suggest that around 34-43% of the stream DIC is from soil respiration.

### 2.4.6 $\Delta^{14}C$ and $\delta^{13}C$ composition of mesozooplankton

The $^{13}C$-depletion of $\delta^{13}C_{Zoops}$ relative to $\delta^{13}C_{POC}$ in August 2009 (as compared to their similarity in June 2009), $^{14}C$-enrichment of $\Delta^{14}C_{Zoops}$ relative to co-occurring $\Delta^{14}C_{POC}$ (a trend that holds true for both measured and large-volume-blank-adjusted $\Delta^{14}C_{POC}$ values) and the similarity of $\Delta^{14}C_{Zoops}$ and $\Delta^{14}C_{DIC}$ values in June and August 2009 (Table 2-6) indicate that the abundance of phytoplankton in the surface water during stratification affords the mesozooplankton the chance to avoid organic detritus, and to selectively feed on living plankton (Marty and Planas 2008). Since suspended POC consists of a mixture of living plankton and detrital materials such as fecal material, crustacean molts and fine particles (Banse 1977), selective feeding on living planktonic organisms yields $\Delta^{14}C_{Zoops}$ similar to $\Delta^{14}C_{DIC}$. Our $\delta^{13}C_{Zoops}$ and $\Delta^{14}C_{Zoops}$ values in the Lake Superior also indicate that the isotopic composition of POC in the surface waters of the lake cannot be reliably used as a proxy for the isotopic composition of algae or phytoplankton, and that the ratio of algal carbon to total POC should be included in mixing models (McCallister et al. 2008).

### 2.4.7 Sources of POC in Lake Superior as revealed by $\Delta^{14}C$, $\delta^{13}C$, and molar C:N ratios

Suspended POM in Lake Superior appears to result from a combination of autochthonous primary productivity, sedimentary resuspension, and terrestrial OM inputs. The net heterotrophy of temperate oligotrophic Lake Superior as indicated from $CO_2$ efflux estimates (McManus et al. 2003; Russ et al. 2004) indicates a significant spatial or temporal subsidy of either terrigenous or sedimentary OM into the lake water column. Most of the $^{14}C$-depleted (old) carbon in the POC pool is potentially sourced from either older terrestrial organic matter or from the lake sediments whereas the $^{14}C$-enriched (fresh labile) portion of POC appears to come from fresh algae based on the difference in surface water and deep water $\Delta^{14}C_{POC}$ values.

In order to tease apart the possible sources of the suspended POC, we correlated measured $\Delta^{14}C_{POC}$ and POC concentrations (Fig. 2-5A) for all depths and seasons. Our data show that the $\Delta^{14}C_{POC}$ values are more $^{14}C$-depleted ($\leq$ 15‰) at lower (< 4 μmol $L^{-1}$) POC concentrations. Two departures from this trend were the June 2009 OL sample from 30 m depth (the depth of the DCM during August-September) and the August 2009 BR sample; both were $^{14}C$-depleted (-12‰ and 7‰, respectively) at higher POC concentration. These samples most likely contained resuspended organic particles from the sediment. In general, deep water POM was $^{14}C$-depleted relative to other POM samples. The surface water POM, with contributions from fresh algae, zooplankton and bacterial matter was $^{14}C$-enriched, whereas the POM from the DCM-depth reflected an intermediate isotopic signal relative to the deep and surface waters.

A cross-plot of measured $\Delta^{14}C_{POC}$ and $\delta^{13}C_{POC}$ clusters the POC (Fig. 2-5B), separating surface-water samples from the stratified season, based upon less $^{13}C$-depleted POC (most likely due to a drawdown of available inorganic carbon due to increased primary production) and more modern $\Delta^{14}C$ values, again recording the DIC signal. Deep-water samples from the stratified period are more depleted in both $^{13}C$ and $^{14}C$, indicating the effects of diagenesis and perhaps sediment resuspension, while DCM-depth samples showed similar $\Delta^{14}C$ to surface samples but more $^{13}C$-depleted signatures, perhaps due to low algal growth rates in the cooler deeper DCM. Samples during the isothermal period are, in general, more depleted in $^{13}C$ than those from stratified periods, and have variable $\Delta^{14}C$ that does not appear correlated with depth. We interpret this as showing the effects of detrital inputs and sediment resuspension throughout the dynamic, mixed water column.

Our estimates of the proportion of terrigenous carbon to the suspended POC pool are consistent with values obtained from the branched and isoprenoid tetraether (BIT) index of suspended POC at the same sampling site and depths in Lake Superior (M. Woltering et al. unpubl.). The BIT index is a measure of the ratio of aquatic vs. terrestrial glycerol dialkyl glycerol tetraether (GDGT) membrane lipids, and is generally used to determine relative amounts of terrigenous organic carbon in paleo-climate (e.g., past temperature) reconstructions (Hopmans et al. 2004). The high percentage of terrestrial OC observed in the stratified deep waters is reflective of low algal production in the deep waters. The observed positive relationship between estimated terrestrial carbon input and $\Delta^{14}C_{POC}$ (Fig. 2-5C) suggests that most of the terrestrial carbon subsidies to the POC pool are $^{14}C$-enriched (modern).

## 2.4.8 Sources of DOC in Lake Superior as revealed by $\Delta^{14}C$ and $\delta^{13}C$

For DOC, the potential sources in Lake Superior are phytoplankton exudation or zooplankton sloppy feeding, desorption from sinking POC, terrestrial inputs from the watershed, and porewater fluxes from the sediment. A plot of water-column $\Delta^{14}C_{DOC}$ vs. DOC concentration (Fig. 2-6A) isolated the stratified-period deep water from the other samples, with this sample exhibiting lower concentration and radiocarbon composition than the other stratified-period samples. The isotopic compositions of DOC in the isothermal water column were intermediate relative to the stratified deep water vs. surface water. Concentrations of DOC during the isothermal period were similar to those in the hypolimnion during stratification. A plot of $\Delta^{14}C_{DOC}$ vs. $\delta^{13}C_{DOC}$ (Fig. 2-6B) again isolated the stratified deep water DOC from the remaining DOC pool (primarily due to radiocarbon 'age'). This separation suggests an older source of DOC to the deep water column, consistent with porewater DOC inputs.

In most cases, pre-aged organic carbon was a considerably higher proportion of the suspended POC than it was in concurrent DOC. This indicates that in Lake Superior, the DOC, which is the largest organic carbon pool in the lake, has more modern, and potentially biogeochemically labile, components, which is consistent with previous studies identifying DOC as critical to both bacterial productivity and respiration in the lake (McManus et al. 2003; Cotner et al. 2004). This observation contrasts with ocean systems (McNichol and Aluwihare 2007), where the bulk DOC pool is dominated by pre-aged organic carbon, but is consistent with river systems (Raymond and Bauer 2001; Raymond et al. 2004) where the bulk POC pool is dominated by pre-aged organic carbon.

## 2.4.9 Biogeochemical carbon cycling insights from $\delta^{13}C$ and $\Delta^{14}C$

The $\delta^{13}C$ and $\Delta^{14}C$ values for DIC in the lake were, as expected, $^{14}C$-enriched relative to the organic matter pools, with $\Delta^{14}C_{DIC}$ values showing slightly more bomb radiocarbon relative to atmospheric $\Delta^{14}C$ at each sampling time. The DIC reservoir appears to cycle rapidly in the lake, on the order of 3-5 years. The depth gradient of $\Delta^{14}C_{DIC}$ minus $\Delta^{14}C_{DOC}$ ($\Delta\Delta^{14}C_{DIC-DOC}$) (Fig. 2-7A) indicates that the DOC pool cycles on a longer time scale than the DIC pool.

The oldest bulk DOC we measured in the lake water column (excluding the anomalous pre-aging in September 2007 which was not replicated in our later samplings) was modern (post 1950), which puts the residence time of the bulk DOC pool at $\leq$ 60 years. This is in general agreement with estimated DOC turnover rates of ~8 years by Urban et al. (2005) and 26-36 years by Cotner et al. (2004) based on input and export carbon fluxes and the size of the DOC pool. Variations in DOC radiocarbon content with depth and time of year indicate, unsurprisingly, that bulk DOC consists of multiple pools of varying lability, and that the most refractory component appears to cycle within the water column for some decades.

The oldest POC we measured in the western basin of the lake was aged (85 ± 25 years BP) to about half the hydrologic residence time of the lake, though blank-adjustment (if appropriate) shifts these values to slightly post-bomb (indicating incorporation of carbon from within the last 60 years). The depth gradient of $\Delta\Delta^{14}C_{DIC\text{-}POC}$ ($\Delta^{14}C_{DIC}$ minus measured $\Delta^{14}C_{POC}$) (Fig. 2-7B) in the thermally stratified lake suggests that the suspended POC pool recycles slowly relative to water-column mixing, and on a somewhat longer time scale than the bulk DOC. This contrasts with observations in open Atlantic and north central Pacific oceans in which the suspended POC pool cycled on much shorter time scales (years to decades) relative to the DOC pool (centuries to millennia, Druffel and Williams 1990; Bauer et al. 2001).

The observed differences in Lake Superior POC vs. DOC cycling relative to the oceans are probably due in large part to the difference in water-column physical parameters in these two systems, where the lake exhibits mixing of the entire water column twice a year. Thus in the lake system, sediment resuspension is more likely to sporadically affect surface and mid-layer waters as well as the deep water-column (adding to suspended POC age), while DOC is mixed into the photochemically and microbially-active surface water semi-annually (thus enhancing water-column DOC removal and keeping the bulk DOC pool semi-labile).

It is worth noting that water-column bulk POC and bulk DOC contain a variety of biochemical compounds (carbohydrates, proteins, lipids, and nucleic acids) that reflect multiple sources, and different pathways of production and decomposition (Hwang and Druffel 2003; Roland et al. 2008). In ocean waters, compound-classes isolated from bulk POC and ultrafiltered DOC have exhibited radiocarbon ages that are both younger and older than the original bulk organic material (Hwang and Druffel 2003; Loh et al. 2004; Roland et al. 2008). Analyzing the $\Delta^{14}C$ in compound-class samples from Lake Superior would enable more robust interpretations of the various sources and biogeochemical processes.

### 2.4.10 Implications of future anthropogenic increases in carbon and nutrients in Lake Superior

Increasing anthropogenic effects such as nutrient and allochthonous carbon loading from surface run-off and sewage discharge, and increasing atmospheric $CO_2$ within the Laurentian Great Lakes region could play important roles in shifting the dynamics of the carbon cycle in Lake Superior. The relatively pristine and oligotrophic nature of the lake puts its ecosystem under great threat from increased nutrient loading. Lake Superior is generally considered to be phosphorus (P) limited, with considerable excess nitrogen available (Cotner et al. 2004; Sterner et al. 2004). Increased P loading from run-off and human sewage, which has high P-levels (Caraco 1995), could lead to eutrophication (Cotner et al. 2004). The current ratio of DOC to POC could be significantly shifted by increasing POC levels from enhanced productivity, and the role of the microbial loop in carbon processing could thus be diminished by eutrophication.

The global carbon cycle has been linked to changes in climate, and the Laurentian Great Lakes are no exception (Nicholls 1999; Sousounis and Grover 2002). Water temperatures and wind speeds in Lake Superior have increased in the past decade, and ice coverage has decreased during the same period in the lake (Austin and Colman 2007; Bennington et al. in press). The increase in water temperature coupled with longer duration of the stratified season could increase the role of microbial processes in converting recently-synthesized POC into DOC (Jiao et al. 2010). Increasing air temperature could also further accelerate microbial turnover rates of DOC, and decreased ice cover along with longer stratified periods could enhance photodegradation of DOC. Wind speed increases may also play a role in a changing carbon cycle, by leading to increases in mixed layer depths.

This is the first study coupling radiocarbon and stable-carbon isotopes, carbon concentrations, and C:N ratios to provide an understanding of the water-column carbon cycle of a great lake system. Both temporal (stratified vs. mixed-water-column) and spatial (depth in open lake and nearshore vs. offshore) variations were important in constraining sources and processes. Such comprehensive approaches in multiple aquatic systems (with increased spatial and temporal resolution) should provide researchers with a deeper understanding of the basic parameters at work in the aquatic portion of the global carbon cycle.

**Table 2-1. Concentrations of bulk DOC (<GF/F, approximately <0.7 $\mu$m), DIC, POC, PON, and particulate C:N within the water column of Lake Superior covering thermally stratified and isothermal water conditions. Molar C:N of particulate organic matter (POM) is based on the ratio of organic C to organic N. For this and later tables, R1 and R2 refer to replicates 1 and 2 from the same sample site and depth. Errors for DOC and DIC are based upon replicate injections, and errors for POC and PON are based upon analyses of multiple external standards. nd = not determined either because sample was not collected or sample was not measured.**

| Date and depth | DIC | DOC | POC | PON | POM |
|---|---|---|---|---|---|
| | ($\mu$mol C $L^{-1}$) | ($\mu$mol C $L^{-1}$) | ($\mu$mol C $L^{-1}$) | ($\mu$mol N $L^{-1}$) | molar C:N |
| Sep 07-Stratified | | | | | |
| OL 5 m | nd | 129.3 ± 4.9 | 8.7 ± 0.2 | 0.99 ± 0.04 | 8.8 ± 0.4 |
| OL 30 m | " | 143.0 ± 4.0 | 10.6 ± 0.2 | 1.30 ± 0.06 | 8.2 ± 0.4 |
| OL 127 m R1 | " | 113.8 ± 5.0 | 2.5 ± 0.1 | 0.29 ± 0.01 | 8.8 ± 0.4 |
| OL 127 m R2 | " | 115.3 ± 4.8 | 3.1 ± 0.1 | 0.35 ± 0.02 | 8.9 ± 0.4 |
| BR 5 m | " | nd | nd | nd | nd |
| May 08-Isothermal | | | | | |
| OL 5 m | 819 ± 4 | 82.0 ± 2.1 | 6.4 ± 0.1 | 0.80 ± 0.02 | 8.0 ± 0.3 |
| OL 30 m | 815 ± 3 | 81.1 ± 2.0 | 7.7 ± 0.2 | 0.94 ± 0.02 | 8.2 ± 0.3 |
| OL 127 m | 809 ± 2 | 80.2 ± 0.6 | 6.5 ± 0.1 | 0.76 ± 0.02 | 8.5 ± 0.3 |
| BR 5 m | 813 ± 2 | 83.0 ± 0.8 | 9.2 ± 0.2 | 1.14 ± 0.03 | 8.1 ± 0.3 |
| Sep 08-Stratified | | | | | |
| OL 5 m | 802 ± 3 | 102.4 ± 1.9 | 15.86 ± 0.05 | 1.70 ± 0.04 | 9.3 ± 0.2 |
| OL 30 m | 821 ± 3 | 92.5 ± 1.1 | 11.93 ± 0.04 | 1.56 ± 0.03 | 7.7 ± 0.2 |
| OL 127 m | 830 ± 3 | 92.8 ± 1.5 | 4.42 ± 0.01 | 0.49 ± 0.01 | 9.1 ± 0.2 |
| BR 5 m | 806 ± 3 | 106.3 ± 1.1 | 11.48 ± 0.04 | 1.44 ± 0.03 | 7.9 ± 0.2 |
| Jun 09-Isothermal | | | | | |
| OL 5 m | 825 ± 3 | 94.2 ± 1.3 | 4.80 ± 0.03 | 0.66 ± 0.002 | 7.26 ± 0.04 |
| OL 30 m | 830 ± 2 | 93.3 ± 1.3 | 4.80 ± 0.03 | 0.57 ± 0.002 | 8.4 ± 0.1 |
| OL 127 m | 822 ± 2 | 93.3 ± 1.2 | 4.70 ± 0.03 | 0.62 ± 0.002 | 7.5 ± 0.1 |
| BR 5 m | 822 ± 2 | 110.8 ± 0.9 | 6.85 ± 0.04 | 0.72 ± 0.002 | 9.5 ± 0.1 |
| Aug 09-Stratified | | | | | |
| OL 5 m | 807.5 ± 0.4 | 104.0 ± 1.9 | 12.0 ± 0.1 | 1.42 ± 0.002 | 8.4 ± 0.1 |

| OL 30 m | 826 ± 2 | 102.6 ± 2.6 | 9.6 ± 0.1 | 1.25 ± 0.002 | 7.69 ± 0.04 |
|---|---|---|---|---|---|
| OL 127 m | 831 ± 2 | 89.3 ± 1.3 | 2.93 ± 0.02 | 0.34 ± 0.001 | 8.7 ± 0.1 |
| BR 5 m | 789 ± 3 | 98.4 ± 2.7 | 13.4 ± 0.1 | 1.60 ± 0.003 | 8.4 ± 0.1 |

**Table 2-2. The $\delta^{13}C$ compositions within the DIC, DOC, and POC in the water column of Lake Superior. Instrumental precision based on analyses of multiple external standards was 0.15‰.**
**nd = not determined either because sample was not collected or sample was not measured.**

| Date and depth | DIC-$\delta^{13}C$ (‰) | POC-$\delta^{13}C$ (‰) | DOC-$\delta^{13}C$ (‰) |
|---|---|---|---|
| Sep 07- Stratified | | | |
| OL 5 m | 1.2 | -27.2 | -19.1 [b] |
| OL 30 m | 0.9 | -29.5 | -22.1 [b] |
| OL 127 m R1 | 0.3 | -29.5 | -19.3 [b] |
| OL 127 m R2 | 0.4 | -30.0 | -25.3 [b] |
| BR 5 m | 1.2 | -28.1 | nd |
| May 08- Isothermal | | | |
| OL 5 m | 0.2 | -30.3 | -26.4 [b] |
| OL 30 m | 0.5 | -30.9 | -25.7 [b] |
| OL 127 m | 0.5 | -30.3 | -27.5 [b] |
| BR 5 m | 0.5 | -30.6 | -25.4 |
| Sep 08- Stratified | | | |
| OL 5 m | 1.5 | -26.7 | -25.6 |
| OL 30 m | 0.6 | -29.4 | -25.9 |
| OL 127 m | 0.2 | -28.7 | nd [a] |
| BR 5 m | 1.2 | -27.4 | -28.9 [b] |
| Jun 09- Isothermal | | | |
| OL 5 m | 0.4 | -29.2 | -26.5 |
| OL 30 m | 0.3 | -29.4 | -26.4 |
| OL 127 m | 0.3 | -29.9 | -26.7 |
| BR 5 m | 0.3 | -30.1 | -26.1 |
| Aug 09- Stratified | | | |
| OL 5 m | 1.3 | -27.8 | -26.1 |
| OL 30 m | 0.6 | -29.7 | -26.0 |
| OL 127 m | 0.2 | -29.5 | -25.8 |
| BR 5 m | 0.9 | -28.3 | -26.5 |

[a]Sample lost during processing in the field or lab.
[b]Based on measurement of freeze-dried DOC samples.

**Table 2-3. $\Delta^{14}C$ within bulk DIC, DOC, and POC in the water column of Lake Superior. The precisions are based on error of standards or of multiple analyses on a target. The values in parenthesis are radiocarbon ages in B.P. years; all other values are modern (post-1950). nd = not determined either because sample was not collected or sample was not measured.**

| Date and depth | DIC-$\Delta^{14}C$ (‰) | POC-$\Delta^{14}C$ (‰) | DOC-$\Delta^{14}C$ (‰) |
|---|---|---|---|
| Sep 07- Stratified | | | |
| OL 5 m | 83 ± 3 | 61 ± 5 | 73 ± 7 [b] |
| OL 30 m | 76 ± 4 | 49 ± 4 | nd [a] |
| OL 127 m R1 | 74 ± 3 | 12 ± 7 | -77 ± 7[b] (585 ± 60) |
| OL 127 m R2 | 80 ± 3 | 28 ± 6 | nd [a] |
| BR 5 m | 79 ± 3 | 31 ± 4 | nd |
| May 08- Isothermal | | | |
| OL 5 m | 71 ± 4 | 35 ± 10 | 49 ± 4 [b] |
| OL 30 m | 66 ± 2 | 41 ± 6 | 39 ± 4 [b] |
| OL 127 m | 63 ± 2 | 18 ± 3 | 59 ± 4 [b] |
| BR 5 m | 68 ± 3 | 40 ± 5 | 43 ± 4 |
| Sep 08- Stratified | | | |
| OL 5 m | 69 ± 3 | 48 ± 3 | 75 ± 4 |
| OL 30 m | 70 ± 4 | 9 ± 3 | 40 ± 4 |
| OL 127 m | 67 ± 3 | 10 ± 3 | nd [a] |
| BR 5 m | 77 ± 4 | 53 ± 3 | 78 ± 4 |
| Jun 09- Isothermal | | | |
| OL 5 m | 56 ± 4 | 21 ± 4 | 49 ± 4 |
| OL 30 m | 60 ± 4 | -12 ± 4 (45 ± 30) | 40 ± 5 |
| OL 127 m | 58 ± 3 | 12 ± 4 | 42 ± 4 |
| BR 5 m | 54 ± 4 | 14 ± 4 | 38 ± 4 |
| Aug 09- Stratified | | | |
| OL 5 m | 61 ± 4 | 33 ± 4 | 51 ± 4 |
| OL 30 m | 59 ± 4 | 30 ± 4 | 46 ± 5 |
| OL 127 m | 52 ± 3 | -17 ± 4 (85 ± 25) | 14 ± 4 |
| BR 5 m | 60 ± 4 | 7 ± 3 | 74 ± 4 |

[a]Sample lost during processing in the field or lab.

[b]Based on measurement of freeze-dried DOC samples.

**Table 2-4. Sedimentary porewater DOC concentrations, and surface porewater DOC isotopic compositions in Lake Superior. Isotopic compositions were determined on only 0-2 cm sediment slices (which include unconsolidated material, i.e., the flocculant layer) from October 2008, June 2009, and August 2009. DOC errors are based on multiple injections, and isotopic errors are based on instrumental precision. OW = refers to water overlying the sediments and collected within the multicore tube. nd = not determined either because sample was not collected or sample was not measured.**

| Depth (cm) | $\mu$mol C $L^{-1}$ | $\delta^{13}C$ (‰) | $\Delta^{14}C$ (‰) |
|---|---|---|---|
| Oct 08 | | | |
| OW | nd | nd | nd |
| 1-2 | 93 ± 1 | nd | nd |
| 2-3 | 145 ± 2 | nd | nd |
| 3-4 | 355 ± 2 | nd | nd |
| Jun 09 | | | |
| OW | 102 ± 2 | nd | nd |
| 0-2 | 108 ± 2 | -35.6 ± 0.2 | -302 ± 5 |
| 2-4 | 301 ± 3 | nd | nd |
| Aug 09 | | | |
| OW | 118 ± 1 | nd | nd |
| 0-2 | 395 ± 4 | -33.3 ± 0.2 | -158 ± 5 |

**Table 2-5. $\Delta^{14}C$ and $\delta^{13}C$, and concentration of DIC, DOC, and POC in Amity Creek. DIC and DOC errors are based upon replicate injections, and POC error is based on analyses of multiple external standards. Values in parenthesis are radiocarbon ages in B.P years.**

| Carbon pool | Date | Water height (cm) | $\mu$mol C $L^{-1}$ | $\delta^{13}C$ (‰) | $\Delta^{14}C$ (‰) |
|---|---|---|---|---|---|
| Baseflow DIC | 13 Sep 2008 | ~41 | 2421 ± 4 | -10.9 | 21 ± 3 |
| Baseflow DOC | 13 Sep 2008 | ~41 | 489 ± 8 | -28.5 | -13 ± 4 (45 ± 35) |
| Stormflow DOC | 12 Jun 2008 | 53 | 825 ± 8 | -28.1 | 51 ± 4 |
| Stormflow DOC | 12 Jun 2008 | 98 | 885 ± 8 | -28.6 | 62 ± 4 |
| Stormflow POC | 12 Jun 2008 | 53 | 352 ± 1 | -27.3 | 11 ± 2 |

**Table 2-6. $\Delta^{14}C$ and $\delta^{13}C$ within pelagic mesozooplankton. The surface POC and DIC values from Table 3 are included here for comparison to mesozooplankton values in order to assess food sources for mesozooplankton biomass production.**

| | June 2009 | | August 2009 | |
|---|---|---|---|---|
| | $\delta^{13}C$ (‰) | $\Delta^{14}C$ (‰) | $\delta^{13}C$ (‰) | $\Delta^{14}C$ (‰) |
| Surface POC | -29.2 | 21 ± 4 | -27.8 | 33 ± 4 |
| Zooplankton | -29.7 | 57 ± 4 | -31.0 | 62 ± 4 |
| Surface DIC | 0.4 | 56 ± 4 | 1.3 | 61 ± 4 |

**Table 2-7. Results of binary models showing relative contribution of pre-aged and modern carbon to water column POC and DOC, and contribution of terrigenous carbon to POC in Lake Superior. $F_p$ and $F_m$ are fractions in % of pre-aged and modern components, respectively. Average values of DOC and POC replicates (R1 and R2) from 127 m in 2007 were used in the model. *Model could not calculate components. **Not calculated either because sample was not collected or not measured.**

| | POC | | DOC | | Terrigenous POC | |
|---|---|---|---|---|---|---|
| Date and depth | $F_p$ | $F_m$ | $F_p$ | $F_m$ | $\mu$mol C L$^{-1}$ | % |
| Sep 07- Stratified: OL 5 m | 20 | 80 | 9 | 91 | 1.3 | 15 |
| OL 30 m | 31 | 69 | 77 | 23 | 1.2 | 11 |
| OL 127 | 66 | 34 | * | * | 0.5 | 16 |
| BR 5 m | ** | ** | ** | ** | ** | ** |
| May 08- Isothermal | | | | | | |
| OL 5 m | 38 | 62 | 23 | 77 | 0.6 | 10 |
| OL 30 m | 31 | 69 | 33 | 67 | 0.8 | 11 |
| OL 127 m | 56 | 44 | 12 | 88 | 0.9 | 13 |
| BR 5 m | 33 | 67 | 29 | 71 | 0.9 | 10 |
| Sep 08- Stratified | | | | | | |
| OL 5 m | 22 | 78 | * | * | 3.0 | 19 |
| OL 30 m | 65 | 35 | 31 | 69 | 0.9 | 7 |
| OL 127 m | 63 | 37 | 74 | 26 | 0.8 | 17 |
| BR 5 m | 17 | 83 | * | * | 1.1 | 9 |
| Jun 09- Isothermal | | | | | | |
| OL 5 m | 44 | 56 | 9 | 91 | 0.2 | 4 |
| OL 30 m | 85 | 15 | 20 | 80 | 0.6 | 13 |
| OL 127 m | 55 | 45 | 18 | 82 | 0.3 | 6 |
| BR 5 m | 52 | 48 | 22 | 78 | 1.4 | 20 |
| Aug 09- Stratified | | | | | | |
| OL 5 m | 32 | 68 | 11 | 89 | 1.5 | 13 |
| OL 30 m | 36 | 64 | 17 | 83 | 0.7 | 8 |
| OL 127 m | 92 | 8 | 55 | 45 | 0.4 | 15 |
| BR 5 m | 63 | 37 | * | * | 1.6 | 12 |

**Table 2-8. Reported concentrations of DIC, DOC, and POC in offshore Lake Superior water column over the past four decades. Although not exhaustive, this is detailed enough to reflect historical trend in the lake. These are reported averages, and the values in parentheses are reported ranges. For multiple depths, the order of the values represents their corresponding concentrations.**

| Date of sample collection | Depth (m) | Number of samples ($n$) | DIC ($\mu$mol C $L^{-1}$) | DOC ($\mu$mol C $L^{-1}$) | POC ($\mu$mol C $L^{-1}$) | Reference |
|---|---|---|---|---|---|---|
| May-Nov 1973 | ≤ 100 | 120 | | | 4-16 | Halfon 1984 |
| Mar 1975 | < 1 | 9 | 733 (683-825) | 92 (83-208) | | Maier and Swain 1978 |
| Jun-Sep 1983 | 5-10 | 4, 8; DOC,POC | | 210 (158-317) | 9-34 | Baker and Eisenreich 1985 |
| Jul-Aug 1986 | 5-10 | 9,5; DOC,POC | | 133 (94-178) | 18 (10-27) | Baker and Eisenreich 1989 |
| 1990 | | 6 | | | 16.1 | Hecky et al. 1993 |
| Spring 1993 | | | | | 6 | Anderson et al. 1998 |
| Jun 1994 | 5-220 | 16 | | | 1-3 | Ostrom et al. 1998 |
| May-Jul 1998 | 3 | 460, 293; DOC,POC | | 116 | 7 | Urban et al. 2005 |
| Aug 1999 | 1-24 | 3 | 965 | 107 | | Ma and Green 2004 |
| Sep 1999 | 20, 30, 240 | 6 | | 112, 113, 98 | | Biddanda and Cotner 2003 |
| May-Oct 1999 | | | | 127 | | Urban et al. 2005 |
| May-Oct 2000 | | | | 123 | | " |

| | | | | | | |
|---|---|---|---|---|---|---|
| Jul 2002 | 5, 30, 200 | 1 | | | 13, 21, 13 | Cotner et al. 2004 |
| Aug 2006 | 5, 30, 80 | 1 | 804, 775, 794 | 108, 103, 107 | 7, 18, 6 | Sterner 2010 |
| Sep 2006 | 5 | 6 | | 210 | | Minor and Stephens 2008 |
| Jul 2007 | 5, 30, 80 | 1 | 773, 788, 773 | 125, 118, 128 | 10, 11, 6 | Sterner 2010 |
| Sep 2007 | 5, 30, 127 | 9 | | 129, 143, 115 | 9, 11, 3 | This study |
| Nov 2007 | 5, 30, 80 | 1 | 812, 813, 822 | 168, 118, 140 | 7, 7, 4 | Sterner 2010 |
| Apr 2008 | " " " | " | 791, 788, 779 | 163, 135, 126 | 2, 2, 2 | " |
| May 2008 | 5, 30, 127 | 9 | 819, 815, 809 | 82, 81, 80 | 6, 8, 7 | This study |
| Jul 2008 | 5, 30, 80 | 1 | 803, 808, 813 | 88, 84, 81 | 12, 4, 4 | Sterner 2010 |
| Sep 2008 | 5, 30, 127 | 9 | 802, 821, 830 | 102, 93, 93 | 16, 12, 4 | This study |
| Jun 2009 | " " " | 6 | 825, 830, 822 | 94, 93, 93 | 5, 5, 5 | " |
| Aug 2009 | " " " | " | 808, 826, 831 | 104, 103, 89 | 12, 10, 3 | " |

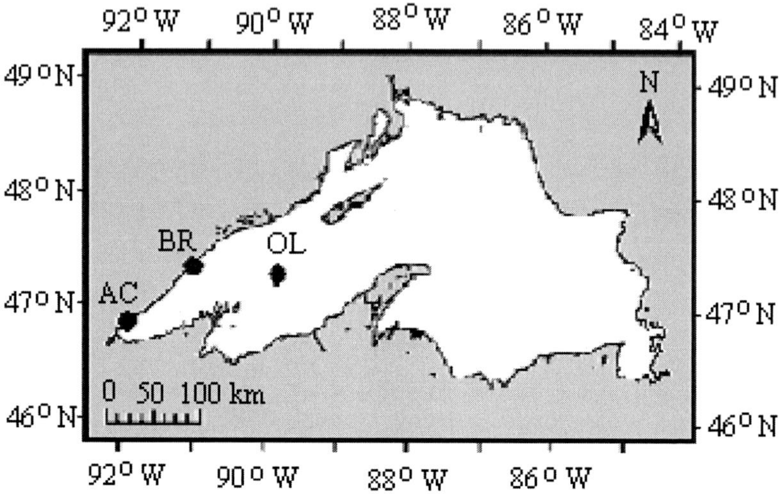


Fig 2-1. Map showing sampling sites: Open-Lake Superior (OL), Baptism River (BR), and Amity Creek (AC).

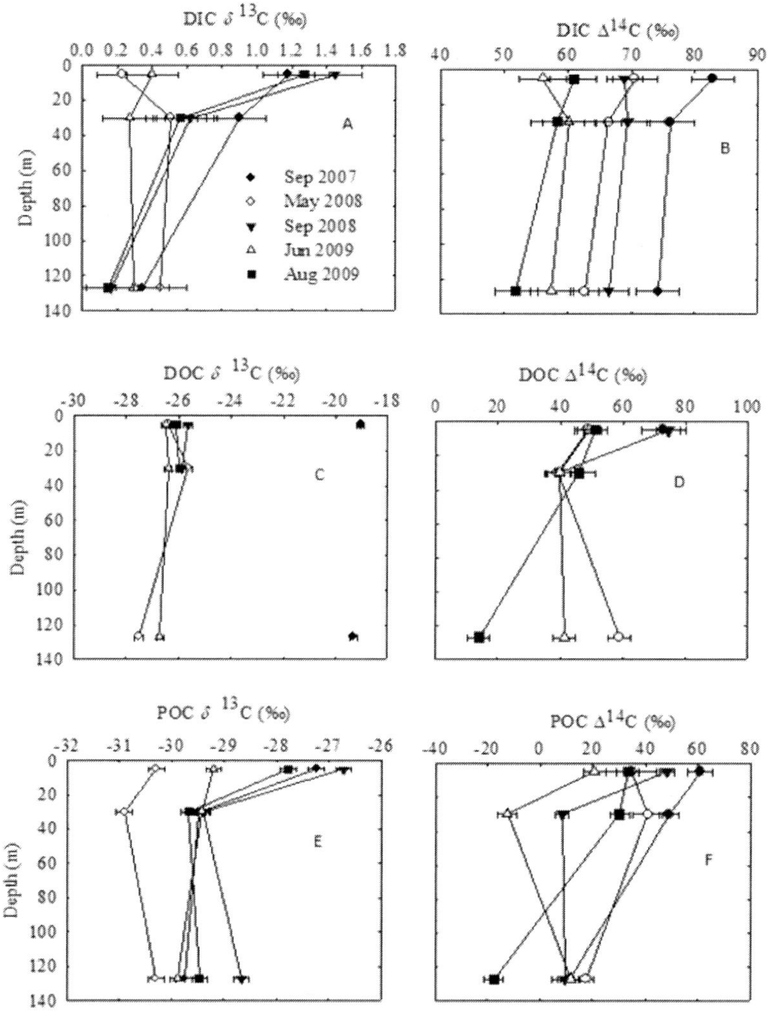


Fig 2-2. The depth profiles of $\delta^{13}C$ within (A) DIC, (C) DOC, and (E) POC, and depth profiles of $\Delta^{14}C$ within (B) DIC, (D) DOC, and (F) POC in the water column of Lake Superior. The data from May 2008 and June 2009 are for isothermal (mixed) condition whereas September 2007, September 2008, and August 2009 are for thermally stratified water column as noted in Table 1. Instrumental error bars are based on analyses of standards or multiple analyses on a target for $\Delta^{14}C$, and analyses of multiple external standards for $\delta^{13}C$.

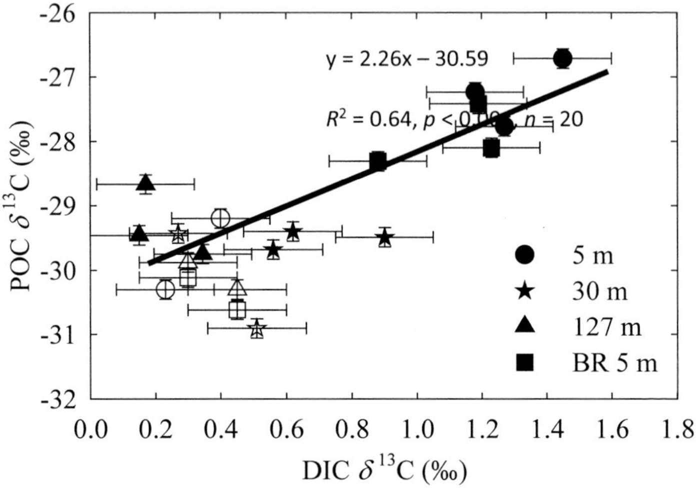


Fig 2-3. Plot of $\delta^{13}C$ in POC vs. DIC. Linear regression shows strong relationship. As is the case for all subsequent plots, the filled symbols are for stratified lake data, and the open symbols are for isothermal data.

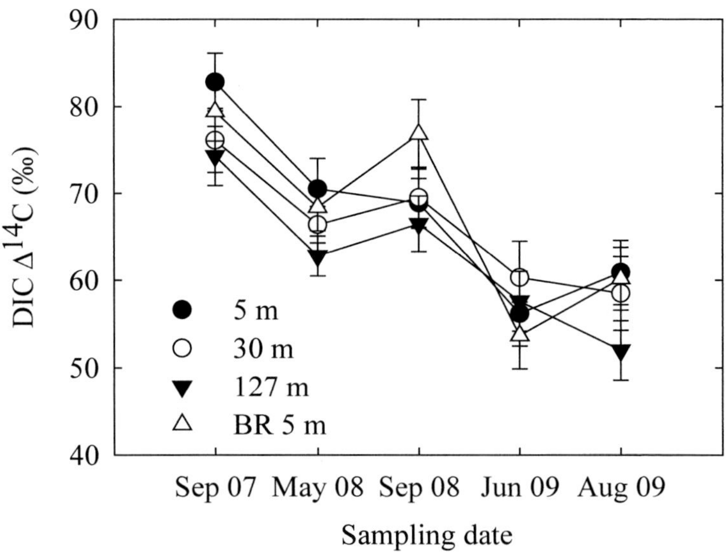


Fig 2-4. The $\Delta^{14}C$ within lake water DIC showing decline over time.

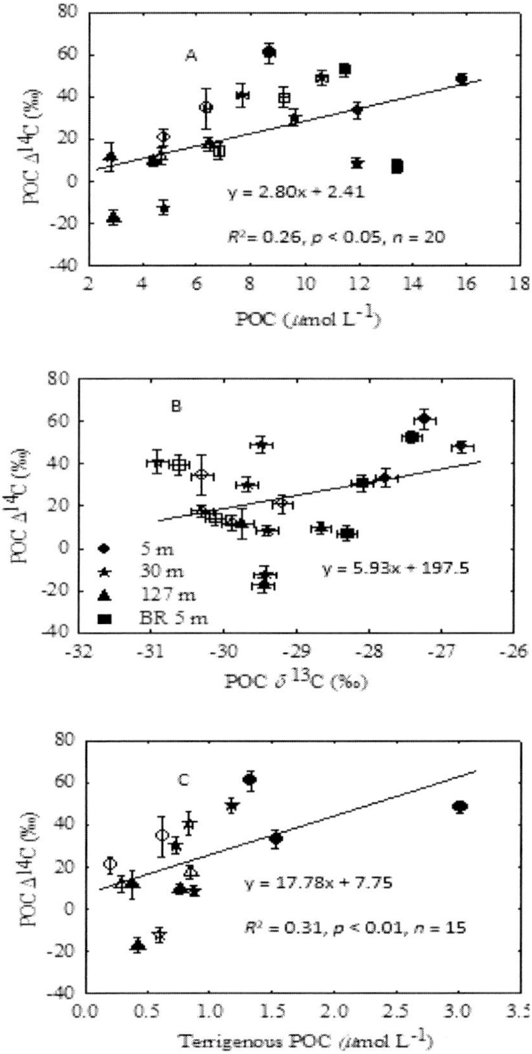


Fig 2-5. The relationship between: (A) $\Delta^{14}C$ and concentration, (B) $\Delta^{14}C$ and $\delta^{13}C$, and (C) $\Delta^{14}C$ and suspended terrigenous POC during thermal stratification (estimated based upon C:N ratios as described in the text) across sampling depth (5-127 m) in the water column of Lake Superior. The plots shows clustering of POC based on potential sources.

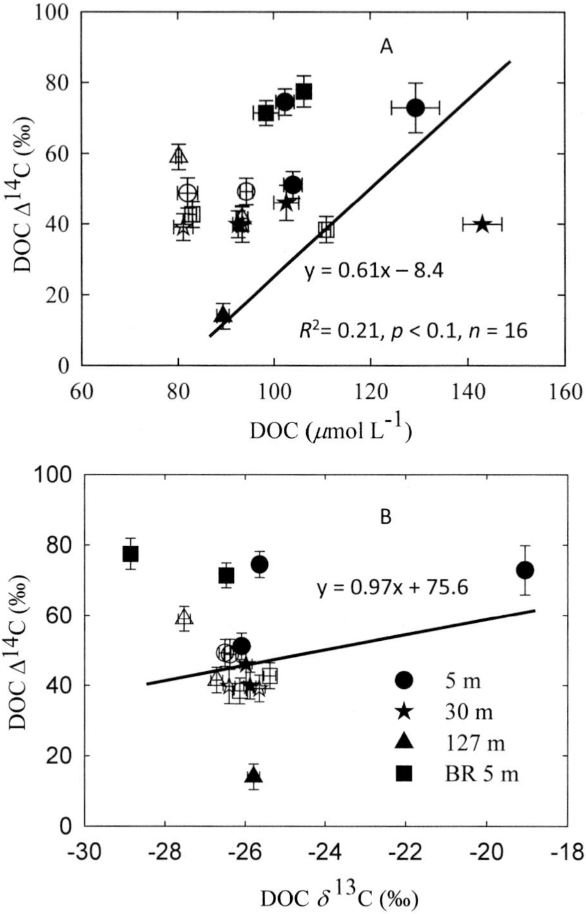


Fig 2-6. Plot of $\Delta^{14}C$ as a function of (A) concentration, and (B) $\delta^{13}C$ within bulk DOC in the water column of Lake Superior. Note that the anomalously $^{14}C$-depleted deep DOC sample ($\Delta^{14}C$ of -77‰) from 2007 was excluded.

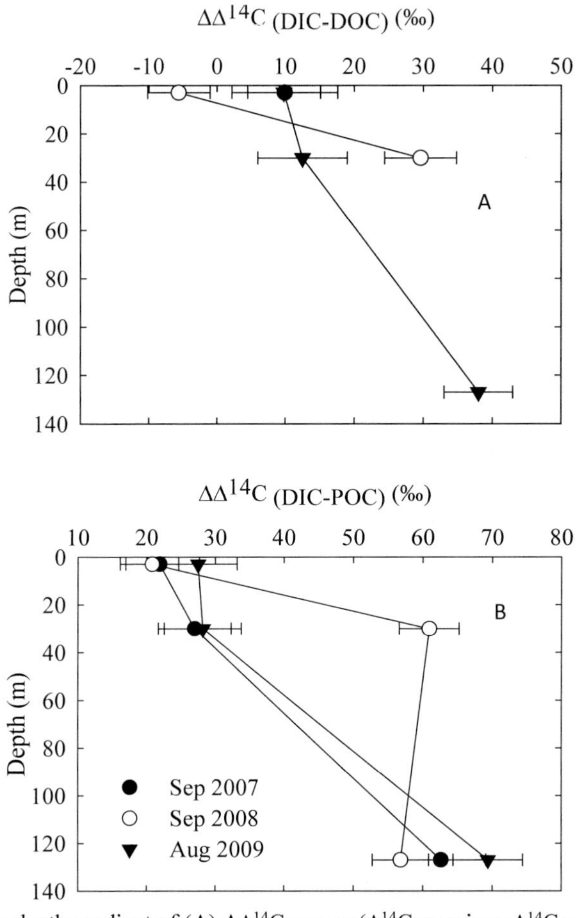


Fig 2-7. The depth gradient of (A) $\Delta\Delta^{14}C_{DIC\text{-}DOC}$ ($\Delta^{14}C_{DIC}$ minus $\Delta^{14}C_{DOC}$), and (B) $\Delta\Delta^{14}C_{DIC\text{-}POC}$ ($\Delta^{14}C_{DIC}$ minus $\Delta^{14}C_{POC}$). These give an indication of relative recycling time and potential deep source (sedimentary input) of organic carbon in the lake. Note that the anomalously $^{14}C$-depleted deep DOC sample ($\Delta^{14}C$ of -77‰) from 2007 was excluded.

# CHAPTER THREE

# RADIOCARBON AND STABLE-ISOTOPE GEOCHEMISTRY OF ORGANIC AND INORGANIC CARBON IN LAKE SUPERIOR

## 3.1 Introduction

Globally, lakes play important roles in the carbon cycle through carbon sequestration in sediments, and efflux of $CO_2$ to the atmosphere (Cole et al. 1994, Cole et al. 2007). The importance of lakes as hotspots of carbon burial is shown by a comparison of average annual carbon burial rates in lakes (on a global basis) vs. the ocean; lakes bury ~25-58% of the carbon that oceans do per year even though the combined surface area of lakes is less than 2% that of the ocean (Dean and Gorham 1998; Cole et al. 2007). Because most lakes are net heterotrophic due to carbon subsidies from the surrounding landscape, they are also hotspots of carbon dioxide efflux, with an estimated potential global $CO_2$ emission of 140 Tg C $y^{-1}$, roughly half the annual carbon transport from rivers to the ocean (Cole et al. 1994). Large lakes such as Lake Superior (with surface areas >10,000 $m^2$) generally have a higher balance of autochthony to allochthony (Kelly et al. 2001), but are also important in sending significant amounts of carbon (including terrestrially-derived carbon) to the atmosphere (Alin and Johnson 2007), and can influence regional and global climate both through this efflux and through the lakes' roles in heat exchange with the atmosphere. Despite the ecological and climatic importance of large lakes, both to their watershed regions and on a global scale, existing data on carbon cycling within them is sparse.

Lake Superior, the earth's largest freshwater lake by surface area, is situated within the North American Precambrian Shield. The maximum and mean depths of the lake are respectively, 406 m and 150 m (Urban et al. 2005), and it is the coldest and deepest of the Laurentian Great Lakes system in the United States and Canada. The pelagic region of the lake is oligotrophic and compares well with open oceanic regions in terms of low organic carbon concentrations, little terrestrial influence, and the dominance of microbial food web in carbon processing (Cotner et al. 2004). Lake Superior therefore serves as an interesting "intermediate" system for comparison with both smaller lakes and larger oceanic systems. It is small enough to study at a basin wide level, which is logistically difficult with oceanic systems, but large enough to have biogeochemical similarities (low organic carbon concentrations, little terrestrial influence, and the dominance of microbial food web in carbon processing) to oceanic systems. However, unlike oceanic systems, the dimictic (that is, semiannual complete vertical mixing) nature of Lake Superior allows for studying reactions between carbon reservoirs on short time scales (annual-decadal); the view of such reactions in ocean systems is obscured by the long ventilation times and long thermohaline circulation cycles in the oceans. As compared to small lakes, Lake Superior provides a key reference point in the potential continuum of allochthony-to-autochthony ratios hypothesized to be a function of lake size or variables that co-vary with lake size (Kelly et al. 2001).

Lake Superior appears to be an important regional source of $CO_2$ to the atmosphere, with efflux estimates of 3-12 Tg C $yr^{-1}$ (Urban et al. 2005; Alin and Johnson 2007). Current estimates of community respiration of 13-84 Tg C $yr^{-1}$ in the lake far exceed the known organic carbon inputs (2.4-9 Tg C $yr^{-1}$) from atmospheric deposition, riverine flux and autochthonous primary production (Cotner et al. 2004; Urban et al. 2005). This mismatch in organic carbon input and output terms represent a major gap in our present knowledge of the dynamics of the lake ecosystem (Sterner 2010), and could be due to difficulty in constraining the magnitude of the various carbon fluxes, or could indicate that the organic carbon budget is not in equilibrium. In this study we apply radiocarbon and stable C and N isotopes to provide additional information on the lake's C cycle.

In the past few decades, radiocarbon ($^{14}C$) has been useful in assessing the ages, reactivity and sources of carbon in surficial environments. $^{14}C$, the radioisotope of carbon, has a half-life of 5730 years. Therefore, ancient organic material that is older that ~60,000 years or new organic material synthesized from an ancient inorganic carbon source would be $^{14}C$-dead, and will contain essentially no $^{14}C$ (Petsch et al. 2001; McNichol and Aluwihare 2007). Above-ground nuclear testing in the 1950s added a "bomb spike" of radiocarbon to the Earth's C cycle, thus extending radiocarbon usefulness to modern/decadal-level environmental processes (described in detail in (McNichol and Aluwihare 2007). Investigating the distribution of $^{14}C$ in the natural environment has provided valuable insights into the sources and/or fate of carbon in the ocean water column (Bauer and Druffel 1998; McCarthy et al. 2011; Pohlman et al. 2011), marine sediments (Eglinton et al. 1997; Goni et al. 1997), and the atmosphere (Hsueh et al. 2007; Rose et al. 2010). Natural $^{14}C$ activity has also been useful in characterizing processes and/or sources of carbon in rivers (Raymond and Bauer 2001), lakes (McCallister and del Giorgio 2008), groundwater (Murthy et al. 1985), rainwater (Raymond 2005), and soils (Trumbore 1997; Petsch et al. 2001). However, these studies have been limited to discrete observations or comparisons within small sample sets taken across space or time.

Our objectives in the present study were to investigate lateral heterogeneity in the sources and biogeochemistry of carbon in Lake Superior using carbon concentration and isotopic distributions, and the effect of thermal stratification on the dynamics of mean ages and/or reactivity and origins of organic carbon across the lake. The Lake Superior data set collected in this study thus represents the first lake-wide radiocarbon study in a large lake system and one of the few studies in any aquatic system to address temporal as well as spatial variations in the radiocarbon composition of the major water-column carbon pools. We measured the natural abundance isotopic ($\Delta^{14}C$ and $\delta^{13}C$) composition and concentrations of particulate organic carbon (POC), dissolved organic carbon (DOC), and dissolved inorganic carbon (DIC) at nearshore and offshore sites in Lake Superior. Together, the POC, DOC, and DIC constitute the total carbon reservoir in the lake water column as the lake is undersaturated with respect to calcium carbonate. Due to the cost and effort currently associated with radiocarbon measurements, the spatial and temporal coverage here is limited to eight sampling sites visited twice and measured at two to three depths on each visit. While limited in spatial and temporal coverage, this sampling scheme represents one of the most extensive aquatic-system radiocarbon studies performed thus far. The resulting data reveals large-scale variations within Lake Superior's carbon pools and give an idea of the temporal variability that may occur in the system. Further, more-detailed temporal and spatial sampling would considerably refine this initial isotopic view of the lake's carbon cycle.

# 3. 2 Methods

## 3.2.1 Sampling

Cruises were undertaken on the R/V Blue Heron to sample the thermally stratified water column in August and the isothermal (mixed) water column in June in 2009. Site locations, water depths, and sampling depths are given in Table 1 and Fig 1. At each of the open-lake (OL) sites, we sampled at 3 depths corresponding to the epilimnion or surface (5 m), the location of the summer deep chlorophyll maximum (DCM, 30-40 m), and the hypolimnion or deep water (127-340 m). At the nearshore (NS) sites, the depths were selected to reflect surface and underlying waters.

At each site, we first obtained temperature, chlorophyll and depth profiles using a Seabird model 911 plus Conductivity, Temperature, and Depth (CTD) rosette equipped with fluorometer, transmissometer, dissolved oxygen sensor, and altimeter. Water samples (Supplementary Fig. 1) were drawn from selected depths using 12 8 L Niskin bottles mounted on the CTD rosette. DIC samples were taken directly from the Niskin bottles via pre-cleaned (10% HCl v/v, and ultra pure water [Millipore Milli-Q Plus]) silicone tubing into previously acid-cleaned and combusted (450°C for 4 hours) 500 mL amber Pyrex bottles. The bottles for DIC were rinsed three times with sample and then overflowed with two volumes of the unfiltered water. After creating a 5 mL headspace, the samples were immediately preserved with saturated mercuric chloride solution, sealed air-tight with glass stoppers coated with Apiezon grease (M), and stored at room temperature in the dark.

Water samples for POC (the fraction that collects on a 0.7 μm, GF/F glass fiber filter) and DOC (the fraction passing through GF/F glass fiber filter) were filtered through pre-combusted Whatman GF/F glass fiber filters (450°C for 4 hours; 0.7 μm nominal pore size) via nitrogen pressurized stainless-steel canisters. For each DOC sample for radiocarbon analysis, approximately 1 L of the resulting filtrate was collected into an acid-washed and combusted glass bottle. Also, approximately 40 mL of the filtrate was collected into an amber glass vial (previously acid-leached and combusted) for DOC concentration measurement. All the DOC samples (both radiocarbon and concentration samples) were preserved by acidifying to pH 2 using 6N HCl (American Chemical Society ACS Plus grade) and subsequent refrigeration. After a known volume (usually about 10 L) of lake water had passed through a GF/F filter, the filter with retained particulate matter was removed from the stainless steel holder, folded, placed in previously-combusted aluminum foil and stored frozen until analysis.

Considerable care was taken throughout sampling to avoid $^{14}C$ contamination. Powder-free nitrile gloves were used during sampling. All plastic tubing (silicone, Teflon, and polypropylene) was cleaned with 10% (v/v) HCl in water, and rinsed with ~10 L of distilled water. All other plastic ware was cleaned with soap and distilled water, leached with 10% (v/v) HCl in water, and then rinsed with distilled water. Glassware was cleaned in the same manner, followed by combustion at 450°C for 4 hours. The R/V Blue Heron and organic geochemistry lab spaces were swiped for possible radiocarbon contamination prior to and during this study and results from both scintillation counting and accelerator mass spectrometry showed no signs of the presence of radiocarbon label.

### 3.2.2 Measurement of concentrations of DOC, DIC, POC, and PON

Concentrations of DOC were measured by high temperature catalytic combustion on a Shimadzu Total Organic Carbon (TOC)-Vcsh analyzer. The TOC analyzer was calibrated with potassium hydrogen phthalate (KHP). Deep seawater from Florida Strait at 700 m (Batch-8 2008) reference waters obtained from the DOC Consensus Reference Program (Hansell laboratory, University of Miami, USA) and additional KHP standards were interspersed and analyzed along with the samples to assess instrumental performance. Our mean DOC value of the external reference sample was 43.9 ± 3.9 µM (n = 5), and agrees with the consensus value of 41-43 µM. For each sample, three injections were performed. If the standard deviation was not ≤ 2.5%, two more injections were performed and the closest three of the five injections were averaged to yield sample concentration (Minor and Stephens 2008).

DIC samples were also analyzed on the Shimadzu TOC-Vcsh Analyzer. In this case, the analyzer was calibrated using primary standard grade sodium carbonate and ACS reagent grade sodium bicarbonate. The inorganic carbon in the sample was volatilized with 25% $H_3PO_4$ (by weight) in a $CO_2$-free closed reaction vessel, and the $CO_2$ evolved was measured by a non-dispersive infrared (NDIR) gas detector as in Zigah et al. (2011). Suspended particulate organic matter (POM) filters were freeze-dried and homogenized, fumigated with 12 N HCl (ACS Plus grade) overnight to remove carbonates, dried again at 60°C, and cooled in a dessicator. They were then loaded into tin capsules and analyzed for particulate organic carbon (POC) and particulate organic nitrogen (PON) concentrations on a Costech ECS 4010 elemental analyzer (EA) coupled to Finnigan Delta Plus XP isotope ratio mass spectrometer (IRMS). The EA was calibrated with acetanilide (Costech Analytical Technologies Inc., Valencia, CA, USA) and caffeine (SigmalUltra; Wheaton Science Products, NJ, USA) reference standard materials, and carbon and nitrogen contents were quantified using the IRMS. Typical instrumental precisions for POC and PON measurements were respectively, 0.2% and 0.4% of the measured concentrations.

### 3.2.3 Black carbon determinations

The black carbon (BC) content of suspended POC was measured via the thermal oxidation method as described in Gustafsson et al. (1997; 2001). Briefly, freeze-dried and homogenized POC filters and NIST Urban Dust standard reference material (NIST SRM 1649a) were loaded into silver cups, placed in a heating block, and set in a combustion oven at 375 °C for 24 hours. The carbon percent was then measured on a Costech ECS 4010 elemental analyzer (EA) as in POC analysis. Our thermal oxidation method yielded a measured value of TOC of 0.1778 ± 0.0001g C $g^{-1}$ dry weight solid and BC of 0.0273 ± 0.0001 g C $g^{-1}$ dry weight solid for the NIST SRM 1649a, similar to reported values of total organic carbon (TOC) of 0.176 ± 0.004 g C $g^{-1}$ dry weight solid and BC of 0.014 ± 0.001 g C $g^{-1}$ dry weight solid (Gustafsson et al. 2001; Reddy et al. 2002).

### 3.2.4 $\delta^{13}C$ and $\delta^{15}N$ measurements

Stable carbon isotope ratios ($\delta^{13}C$) of samples (DOC, DIC, and POC) were determined at the National Ocean Sciences Accelerator Mass Spectrometry Facility (NOSAMS) at the Woods Hole Oceanographic Institution (WHOI) using an Optima IRMS to measure subsamples of the purified $CO_2$ of DOC, DIC, and POC intended for radiocarbon analyses. Stable nitrogen isotope ratios ($\delta^{15}N$) and some additional $\delta^{13}C$ values in POM samples were measured at the Large Lakes Observatory (LLO) using a Finnigan Delta Plus XP IRMS with Conflo III interface (Thermo Fisher Scientific Inc., Waltham, MA) coupled to Costech ECS 4010 EA. The instrument was calibrated with certified reference materials (low organic soil B-2153, sorghum flour B-2159; from Elemental Microanalysis Ltd., Okehampton, Devon, UK) and in-house standards (acetanilide, caffeine, and urea) that have been calibrated against NIST reference standards (NBS

19, NBS 8540, and NBS 8542 for carbon, and NBS 8547, NS 8548, and NBS 8549 for nitrogen. Typical instrumental precision of $\delta^{15}N$ and $\delta^{13}C$ based on analyses of multiple external standards were 0.17‰ and 0.2‰, respectively. The stable isotope ratios ($^{13}C$:$^{12}C$ and $^{15}N$:$^{14}N$) are reported as $\delta^{13}C$ and $\delta^{15}N$ respectively, which are the per mil difference relative to Pee Dee Belemnite carbonate and atmospheric air standards. The δ notation is calculated as: $\delta Q = [R_{sample} / R_{standard} - 1] \times 1000$, where Q is either $^{13}C$ or $^{15}N$, and R is either $^{13}C$:$^{12}C$ or $^{15}N$:$^{14}N$.

### 3.2.5 $\Delta^{14}C$ measurements

Radiocarbon ($\Delta^{14}C$) measurements were performed at NOSAMS. DOC samples were ultraviolet (UV) irradiated using a protocol based upon Beaupre et al. (2007) but using three hours of oxidation instead of four. The evolved $CO_2$ was trapped in a vacuum line, purified cryogenically and reduced to graphite with $H_2$ over Fe catalyst. POC samples were freeze-dried, fumigated with 12 N HCl (ACS Plus grade) for 24 hours to remove carbonates, redried at 60°C, and combusted to $CO_2$ in a modified Carlo Erba NA1500 elemental analyzer. The evolved $CO_2$ was separated cryogenically and reduced to graphite. DIC samples were directly hydrolyzed with $H_3PO_4$, and the resulting $CO_2$ was stripped with nitrogen gas and trapped, cryogenically cleaned, and reduced to graphite as in Zigah et al. (2011).

In all cases, the graphite produced was analyzed by accelerator mass spectrometry (AMS) along with primary and secondary standards, and combustion and graphitization process blanks. Radiocarbon values are reported as $\Delta^{14}C$, the part per thousand deviation of the sample's $^{14}C$:$^{12}C$ ratio relative to a nineteenth century wood standard that has been corrected to the activity it would have had in 1950 and a $\delta^{13}C$ of -25‰. $\Delta^{14}C$ was corrected for fractionation using $\delta^{13}C$ of samples according to the convention of Stuiver and Polach (1977). Instrumental precision of the $\Delta^{14}C$ analysis is based on the error of standards or multiple analyses on a target and were 2-5‰ for DIC, 2-5‰ for DOC, and 3-11‰ for POC.

Because of the challenges inherent in sampling for and analyzing natural-abundance radiocarbon, we performed radiocarbon analysis on POC and DOC method blanks. The POC method blank was obtained by filtering > 100 L of Milli-Q water through pre-combusted GF/F filters using our canister set-up. This large volume is ten times the volume of lake water filtered for POC samples and was designed to provide enough carbon for a radiocarbon measurement. For comparison, we also measured carbon content of a clean GF/F filter. A 1 L portion of the GF/F filtrate was collected as the DOC method blank. Details concerning the POC blank are described in Zigah et al. (2011) and we report both uncorrected (measured) and blank-corrected $\Delta^{14}C_{POC}$ values. The POC-blank $\Delta^{14}C$ value does not affect basic depth and locational trends in the samples discussed here. We have since analyzed the DOC method blank, which generated 8.6 µmol C $L^{-1}$. Just as in the POC method blank (Zigah et al. 2011) the DOC method blank carbon would also include contributions from the MilliQ water as well as the filtration apparatus, and would thus contribute < 9.8 % of the carbon in the least concentrated DOC sample. As the DOC method blank's $\Delta^{14}C$ and $\delta^{13}C$ values were -386 ± 3‰ and -30.4‰, respectively, blank-corrected DOC samples would be shifted by ~38‰ toward more $^{14}C$-enriched values. We report both uncorrected (measured) and blank-corrected values of $\Delta^{14}C_{DOC}$ here as we cannot isolate the MilliQ water contribution (which is not relevant when lake water is filtered) from the filter apparatus contribution, and suspect that much of the carbon in the DOC blank actually originated from the starting MilliQ water. Again, as with POC (Zigah et al. 2011), depth and locational trends remain the same in blank-corrected vs. uncorrected samples.

### 3.2.6 Statistical analyses

We used STATISTICA (StatSoft, Tulsa, Oklahoma, USA) or Excel (Microsoft, USA) in all analyses. The relationships between delta values and/or concentrations of the different carbon pools were assessed with correlation analyses in which we report Pearson's correlation coefficient (R), the probability (P) and the number of samples (n). Statistical significance was tested at the 95% confidence level, thus $\alpha = 0.05$. We checked all data for normality, and those that were not normally distributed (POC and PON concentrations and $\delta^{13}C_{DIC}$) were log transformed before analysis. DIC and DOC concentrations, and translated (thus a constant added to remove negative values) values of $\Delta^{14}C$-POC remained skewed even after log transformation and were not used for statistical analysis. In order to improve statistical power in all analysis, sample sizes less than 30 ($n < 30$) were not used for significance testing.

## 3.3 Results

### 3.3.1 Concentrations of carbon species, particulate nitrogen, and particulate black carbon

The DIC concentration in the lake ranged from 789-871 μM (Fig. 3-2A, B) with the most frequently observed values ranging from 820 to 830 μM (Fig 3-3A). DOC and POC concentrations ranged from 89 to 208 μM (Fig. 3-2C, D) and 2.3 to16.5 μM (Fig. 3-2E, F), respectively, and their frequency distributions show that 80 to 100 μM DOC (Fig. 3-3B), and 4 to 6 μM POC (Fig. 3-3C), were the most commonly observed values. The Lake Superior water column undergoes physical homogenization twice each year, during late spring and early winter. Concentrations of DIC, DOC, POC, and PON during spring mixing in June were uniform vertically (i.e., with depth) and, in general, laterally across the lake, although the nearshore sites (BR and ONT) were higher in concentrations in all the carbon species than the offshore sites (Fig 3-2A-F; Table 3-2). In August when the lake is thermally stratified, the offshore DIC concentration increased with depth from an average of 815 μM in the epilimnion to 826 μM in the deep hypolimnion (Fig. 3-2B). Epilimnetic DOC concentrations were higher during stratification in August than in June as would be expected in the warmer and more productive stratified surface waters. DOC concentrations were also higher (124-208 μM) at the nearshore sites (BR, ONT, NB) than the offshore sites (89-122 μM) during stratification (Fig 3-2D). Of the offshore sites, the highest epilimnetic DOC concentration during stratification was observed in the central province (CM) of the lake. With the exception of the EM site, DOC concentration decreased with depth at the offshore sites during stratification (Fig 3-2D, Table 3-3).

POC and PON concentrations declined with depth during stratification at the offshore sites (Table 3-3). The POM C:N values in the open lake shifted by ~ 0.7 units from a mean of 7.7 (range 7.2-8.4, n = 14 pairs) in the isothermal lake in June to a mean of 8.4 (range 7.6 to 9.2, n = 14 pairs) during stratification in August (Table 3-2 and 3-3). Black carbon concentrations within a subset of suspended POC samples from the isothermal water column in June ranged from 0.17-0.43 μM (mean 0.23 ± 0.09, n = 7) corresponding to mole percent black carbon (BC / POC) of 3.9-8.9% (mean 5.1 ± 1.8%) (Table 3-4).

### 3.3.2 Vertical and lateral variations of stable carbon and nitrogen isotopes

The $\delta^{13}C_{DIC}$ was tightly coupled to the mixing/thermal regime of the lake. During physical homogenization (isothermal condition) in June, the $\delta^{13}C_{DIC}$ values, like DIC concentrations, were spatially (by depth and laterally) uniform, with a mean of 0.4 ± 0.1 ‰ (Fig. 3-4A). However, in the warmer productive surface layer during thermal stratification in August, the $\delta^{13}C_{DIC}$ in the epilimnion averaged 1.0 ± 0.2‰ ($n = 5$), and decreased vertically to an average hypolimnetic $\delta^{13}C_{DIC}$ of 0.2 ± 0.1 ‰ ($n = 5$) (Fig. 3-4B). In the offshore lake, the WM site had the largest shift toward more positive ($^{13}C$-enriched) values in epilimnetic $\delta^{13}C_{DIC}$ after transition from isothermal to stratified conditions. $\delta^{13}C_{DIC}$ was positively and significantly correlated to POC concentration ($R = 0.74$, $P < 0.0001$, $n = 34$), and negatively and significantly correlated to $\delta^{15}N_{PON}$ ($R = 0.68$, $P < 0.0001$, $n = 34$).

The $\delta^{13}C_{DOC}$ values were not coupled to the physical and/or thermal regime of the lake, and did not exhibit considerable vertical or lateral variation. The apparent DCM and epilimnetic $^{13}C$-enrichment of approximately 0.2‰ relative to hypolimnetic $\delta^{13}C_{DOC}$ in June and August is within the precision of $\delta^{13}C$ measurement, except at CM in August where the epilimnion was $^{13}C$-enriched by 2.3‰ relative to the hypolimnion. The surface-to-deep water integrated mean value of $\delta^{13}C_{DOC}$ in the offshore region of the lake was -26.3 ± 0.2‰ in June and -25.8 ± 0.4‰ in August (Fig. 3-4C, D). On the other hand, the $\delta^{13}C_{POC}$ in the epilimnion was responsive to physical regime. Epilimnetic $\delta^{13}C_{POC}$ in the isothermal lake shifted by 1.5 ± 0.6‰ toward more $^{13}C$-enriched values during stratification (Fig. 3-4E, F). There was a lateral gradient in the difference in epilimnetic $\delta^{13}C_{POC}$ between stratified and isothermal conditions ($\Delta\delta^{13}C_{POC}$). $\Delta\delta^{13}C_{POC}$, which is defined here as the stratified $\delta^{13}C_{POC}$ minus isothermal $\delta^{13}C_{POC}$, was higher (~2.2 ‰) at the nearshore sites (BR and ONT). In the offshore sites, the western and southern provinces shifted most, with a $\Delta\delta^{13}C_{POC}$ of 1.4‰ and 1.7‰, respectively (Fig. 3-4E, F).

The $\delta^{15}N_{PON}$ was sensitive to the physical regime of the lake. Epilimnetic $\delta^{15}N_{PON}$ shifted from $^{15}N$-enriched values (range -0.6‰ to 1.7‰; mean 0.5‰) in the isothermal lake in June to $^{15}N$-depleted values (range -3.9‰ to -2.3‰; mean -3.0‰) in the stratified lake in August (Fig. 3-4G, H). An interesting, although not surprising, observation was the consistent $^{15}N$-enrichment of PON with depth in all the offshore sites of the lake during stratification (Fig. 3-4H, Table 3-3). In addition to the high sensitivity of $\delta^{15}N_{PON}$ to the physical regime of the lake, $\delta^{15}N_{PON}$ showed significant and negative correlations with concentrations of POC ($R = -0.71$, $P < 0.0001$, $n = 39$), and PON ($R = -0.71$, $P < 0.0001$, $n = 39$).

### 3.3.3 Distributions of $\Delta^{14}C$ of DIC, DOC and POC

The $\Delta^{14}C_{DIC}$ in the surface-to-deep offshore sites averaged 59‰ (range 52‰ to 63‰, n = 15) in June, and 59‰ (range 50‰ to 65‰, n = 15) in August (Fig 3-5A, B). Distribution of $\Delta^{14}C_{DIC}$ with depth, and with season (isothermal vs. stratified) was spatially homogeneous at the current precision of accelerator mass spectrometry (AMS) of 2-7‰ (Turnbull et al. 2006; McNichol and Aluwihare 2007). At the nearshore sites, $\Delta^{14}C_{DIC}$ ranged from 38‰ off the Ontonagon river (ONT site) to 54‰ off the Baptism river (BR) in June, and averaged 53‰ (range 36‰ to 60‰, n = 5) in August (Fig 3-5A, B).

Uncorrected $\Delta^{14}C_{DOC}$ in the offshore provinces of the lake was similar to or somewhat $^{14}C$-depleted relative to concurrent DIC, with a mean $\Delta^{14}C_{DOC}$ of 41‰ (range 22‰ to 58‰, n = 15) in June, and mean of 40‰ (range 14‰ to 54‰, n = 15) in August (Fig. 3-8A-D). Blank-corrected $\Delta^{14}C_{DOC}$, which may be overcorrected if MilliQ water carbon is a significant portion of the blank as we suspect, was consistently higher than concurrent $\Delta^{14}C_{DIC}$, and ranged from 63-104‰ (n = 15) in June and 52-101‰ (n = 15) in August (Table 3-2, 3-3). An interesting observation in the open lake was the approximately 28 to 30‰ (18-49‰ for blank-corrected values; Fig. 3-5C, D) $^{14}C$-depletion of the epilimnetic DOC from southern, eastern, and northern regions (SM, EM, and NM) in June and SM and NM regions of the lake in August, compared to the corresponding epilimnetic DOC from the other offshore regions of the lake (Fig. 3-5C, D, Table 3-2, 3-3). A depth gradient was observed at the western, central, and eastern provinces of the lake during stratification in August (both for measured, i.e. uncorrected, and blank-corrected $\Delta^{14}C_{DOC}$), with the epilimnetic DOC being $^{14}C$-enriched relative to the hypolimnetic DOC (Fig. 3-5D, Table 3-3). However, an intriguing contrast was observed at the southern and northern provinces of the lake (based on both measured and blank-corrected $\Delta^{14}C_{DOC}$), where the hypolimnetic DOC was $^{14}C$-enriched compared to the epilimnetic DOC (Fig. 3-5D, Table 3-3). At the nearshore sites (BR and ONT), measured $\Delta^{14}C_{DOC}$ was more variable, ranging from -10‰ to 74‰ (mean 40‰, n = 8) (Fig. 3-5C, D), with blank-corrected values ranging from 17-94‰ (Table 3-2, 3-3). The most

$^{14}$C-depleted offshore DOC was seen in the deep water at the western province (measured $\Delta^{14}$C = 14‰) during stratification in August (Fig. 3-5D), and the most depleted (and actually pre-bomb) nearshore DOC was observed at the ONT site (measured $\Delta^{14}$C = -10‰) in June (Fig. 3-5C). These values are shifted to more $^{14}$C-enriched values when blank-corrected as shown in Table 3-2 and Table 3-3.

In most cases, POC samples were more $^{14}$C-depleted than co-occurring DOC samples. Measured $\Delta^{14}$C of POC in the isothermal open lake varied widely between 34‰ (68‰ for blank-corrected values) and -303‰ (-354‰ for blank-corrected values), corresponding to 'modern' and 2840 yrs B.P (3453 yrs B.P. for blank-corrected values), respectively (Fig. 3-5E, Table 3-2). Variation in $\Delta^{14}$C was similarly wide in open lake POC from August, with measured and blank-corrected values, respectively, from 58 to 75‰ (modern) to -145‰ to -170‰ (1200-1440 yrs B.P) (Fig. 3-5F, Table 3-3). While $^{14}$C-depletion was seen across both surface and deep POC samples during isothermal conditions in June, the $^{14}$C-depleted samples were mostly limited to deep-water POC during stratification in August (Fig. 3-5E, F). The measured $\Delta^{14}C_{POC}$ from the nearshore regions of the lake averaged 13‰ (range -19‰ to 36‰, n = 8) (Fig. 3-5E, F). The oldest offshore POC ($\Delta^{14}$C = -303‰ to -354‰; 2840-3453 yrs B.P. at EM in June, and -145‰ to -170‰; 1200-1440 yrs B.P. at SM in August) were seen in the deepest basins of the lake (Fig. 3-6). Nearshore POC showed much less $^{14}$C-depletion, with the most depleted nearshore sample (NB) having $\Delta^{14}$C = -19‰ to -14‰ (56-95 yrs B.P). Although a considerable number of the open-lake POC samples were old, most (~70%) of the POC samples from the lake had a modern (post-1950) $^{14}$C signal (measured $\Delta^{14}$C of 0-50‰, Fig. 3-7C).

## 3.4 Discussion

### 3.4.1 Physical regime and the distribution of DIC concentration and isotopes

In general, DIC concentrations and $\delta^{13}$C in the open lake appear laterally homogenous with no significant localized influence (Fig. 3-2A, B; 3-4A, B) and with their dynamics modulated by, and tightly coupled to the physical homogenization of the water column in June (isothermal conditions). Photosynthesis and respiration appear responsible for the open-lake depth trends in August, where the epilimnion is lower in DIC concentrations and has a higher $\delta^{13}C_{DIC}$, due to photosynthetic depletion and fractionation of the DIC pool (coupled with lower DIC solubility in warmer waters), while the hypolimnion shows higher DIC concentrations and lower $\delta^{13}C_{DIC}$, indicating the effects of increased remineralization of organic carbon at depth (Velinsky and Fogel 1999; Zigah et al. 2011). The largest shift of $\delta^{13}C_{DIC}$ between isothermal and stratified conditions was observed in the epilimnion at the western province (WM) and could be due to the higher primary productivity reported for this region of the lake relative to the central lake and eastern basin (McManus et al. 2003; Cotner et al. 2004). The general homogeneity of $\Delta^{14}C_{DIC}$ across the entire open lake is consistent with the lake's large surface area, spatially well-mixed water column, and the absence of significant localized groundwater input or input of $^{14}C$-dead carbon from dolomite or carbonate. Values of $\Delta^{14}C_{DIC}$ (50-65‰) in the open lake were higher than that of co-occurring atmospheric $CO_2$ (38 ± 2‰; Zigah et al. 2011) as determined from the $\Delta^{14}C$ content of leaves of a local corn plant collected the same year (Hsueh et al. 2007). This offset is most likely due to the time of exchange for the in-lake DIC pool with the atmosphere, which is affected by water-column stratification, ice cover, and the extent of water column mixing (Hesshaimer et al. 1994; Abbott and Stanfford 1996; Hughen et al. 2004; Urban et al. 2005; Alin and Johnson 2007). In the nearshore regions, the low values of $\Delta^{14}C_{DIC}$ observed off the Ontonagon ($\Delta^{14}C_{DIC}$ = 38‰) in June and in Nipigon Bay ($\Delta^{14}C_{DIC}$ = 36‰) in August were similar to the atmospheric $\Delta^{14}C$, and suggest mineralization of recent terrestrially-derived OM (such as leaf material) and/or quicker equilibration of these shallower near-shore sites with the overlying atmosphere.

### 3.4.2 DOC biogeochemistry

Vertical distributions of DOC concentration and $\delta^{13}C_{DOC}$ in June, which were homogeneous (Fig 3-2C), are consistent with the physical regime of the lake, which exhibits complete water column mixing at this time of the year. However, the lateral similarities in concentration and stable isotope composition in June are a bit more surprising as modeling studies have shown exchange between basins on the order of months to years (personal communications, V. Bennington and P. Cheng) and differ from $\Delta^{14}C_{DOC}$ results indicated below, perhaps because radiocarbon presents a larger dynamic range ($\Delta^{14}C$, -1000‰ to +200‰) and is therefore more sensitive to variations in the carbon cycle than are concentrations or $\delta^{13}C$ values ($\delta^{13}C$ for organic carbon range from -32 to -12‰) (Petsch et al. 2001; Wakeham et al. 2006).

Heterogeneity in DOC concentration during stratified conditions in August (Fig. 3-2D) is most likely due to differences in productivity in the epilimnion and heterotrophy in the deeper waters. The western province of the lake has been shown to have higher primary production relative to other parts of the lake (McManus et al. 2003; Cotner et al. 2004), consistent with the relatively higher DOC concentration at WM relative to EM, and with the trends in DIC concentration and isotopic composition discussed in the previous section. Site CM can be impacted by the Keweenaw current (Zhu et al. 2001) and sites SM and NM by the generally cyclonic gyre circulation in the eastern basin (Bennington et al. 2010) thus these locations are affected by water from nearshore regions, with its higher concentrations of land-derived material, nutrients, and warmer summer temperatures, the latter two of which could enhance primary productivity in August. The $\delta^{13}C$ of DOC in most cases exhibited no significant variation laterally or vertically in August within the instrumental precision of 0.2‰ (Table 3-3), possibly because water-column mixing (twice per year) and lateral flow distributes it more quickly than biogeochemical processes can inflict a fingerprint upon it. The $^{13}C$-enrichment with depth seen at CM and WM during stratification could be due to diagenetic alterations or remineralization in the water column (Vahatalo and Wetzel 2008).

Throughout the lake, both nearshore and offshore, the bulk $\Delta^{14}C_{DOC}$ (both measured and blank-corrected) was modern (post-bomb) with the exception of site ONT in June (Fig. 3-5C, D). The $^{14}C$-depletion of the surface DOC at NM, EM, and SM sites during spring mixing in June (relative to WM and CM sites) may be due in part to the current structure of Lake Superior (as described in Bennington et al. 2010), where sites NM and SM, and to some extent EM, are in the flow paths of the generally cyclonic flow in that region of the lake and would be affected by nearshore-derived organic matter entrained into these currents, especially in winter and spring. In summer, sites NM and SM are likely to remain affected by this cyclonic flow and nearshore organic matter transported in it, while site EM is less impacted and shows epilimnion $\Delta^{14}C_{DOC}$ values similar to the central and western-lake sites (Fig. 3-5D). Another process/mechanism that could explain the $^{14}C$-depletion of epilimnetic DOC seen at SM and NM sites is photoproduction of DOC from POC as observed in several natural systems (Komada and Reimers 2001; Koelmans and Prevo 2003; Kieber et al. 2006; Butman et al. 2007; Mayer et al. 2006, 2009; Shank et al. 2011). Also, OC in rainwater could be a source of $^{14}C$-depletion of DOC in the epilimnion as the lake is considerably influenced by its ‘airshed’ due to its large surface area

(Cotner et al. 2004). Old rainwater OC has been observed in Woods Hole, MA and New Haven, CT in USA ($\Delta^{14}C$ = -653‰, 8567 yrs B.P; Raymond 2005) but $\Delta^{14}C$ of rainwater over Lake Superior is not available. We expect the influence of photoproduced DOC and rainwater DOC to be widespread across the lake rather than being localized at NM and SM as was seen in the lake, hence the likely cause of the surface water DOC $^{14}C$-depletion is nearshore old DOC transported by currents as discussed above.

During the stratified period, DOM at the WM, CM, and EM sites shows depletion in $^{14}C$ with depth in the water column suggesting some input of old DOC to the hypolimnetic waters at these sites. Porewater-derived DOC could potentially be the source of this old POC. Also, lateral deep water current could transport old DOC from the nearshore region to the offshore sites, especially at CM site which is close to the nearshore ONT site.

Although DOC in groundwater could also be considerably $^{14}C$-depleted (Murthy et al. 1985), groundwater is not a significant source of deep water in Lake Superior, and likely has very little influence on the lake DOC. This is also confirmed by the absence of considerable $^{14}C$-depletion in our hypolimnetic $\Delta^{14}C_{DIC}$ as the presence of groundwater should also affect lake DIC. In general, the bulk DOC pool appears semi-labile throughout the lake because of the post-bomb $\Delta^{14}C$ values indicating DOC turnover times of $\leq$ 60 years. Using our DOC pool size (calculated from concentration data and lake volume as 14.2-16.4 Tg C) and the DOC influx rate of 0.7-1.3 Tg C $yr^{-1}$ reported by Cotner et al. (2004) puts the DOC residence time at 11-23 years, whereas the DOC influx rate of 0.1 Tg C $yr^{-1}$ reported by Urban et al. (2005) give a DOC residence time of 16-41 years. These turnover times are consistent with previous estimates of approximately one to four decades (Cotner et al. 2004; Urban et al. 2005) based upon carbon budget approaches.

### 3.4.3 Dynamics of C and N concentrations and stable isotope ratios within POM

Higher epilimnetic POC and PON values observed during stratification (Table 3-3) could be due to increased algal, bacterial and zooplankton production (Velinsky and Fogel 1999). The lakewide C:N values of POM agree with previously reported values (C:N values of 4.6-14.3) (Guildford and Hecky 2000; Urban et al. 2004; Sterner 2010; Zigah et al. 2011). The positive shift in the POM C:N values during stratification implies a disproportionately higher increase in organic carbon relative to organic nitrogen (or preferential mineralization of PON) most likely due to shifts in algal and/or bacterial communities and their biomass production vs. remineralization demands.

Increased aquatic in situ primary production in the warmer epilimnetic waters most likely yielded the observed shift in $\delta^{13}C_{POC}$ toward more $^{13}C$-enriched values during stratification (Fig. 3-4D) (Keough et al. 1996, 1998; Sierszen et al. 2006). Just as in $\delta^{13}C_{DIC}$, the largest shift in $\delta^{13}C_{POC}$, occurring at the western province during stratification (Table 3-3), was likely due to relatively higher primary productivity at this site (McManus et al. 2003; Cotner et al. 2004). Our $\delta^{15}N_{PON}$ values are consistent with previously reported values in the lake of -1.0‰ to 3.0‰ in May-June, and -4‰ to 1‰ in August-September (Pang and Nriagu 1977; Keough et al. 1996; Ostrom et al. 1998; Harvey and Kitchell 2000; Sierszen et al. 2006; Kumar et al. 2011). The shift in $\delta^{15}N_{PON}$ to more $^{15}N$-depleted values during stratification is likely due to utilization of more nitrate relative to ammonium by phytoplankton. Nitrate ($NO_3^-$) is the most abundant inorganic nitrogen (N) species in the lake (~25 µM; Sterner et al. 2007; Sterner 2010) with $^{15}N$-depleted values of $\delta^{15}N$ (-2.3 ± 0.3‰; Finlay et al. 2007). On the other hand, ammonium ($NH_4^+$) concentration in the lake is low (~0.2 µM; Sterner 2010) but constitutes a disproportionately larger (~75%) portion of the total N uptake by plankton (Kumar et al. 2008), and has comparatively $^{15}N$-enriched values ($\delta^{15}N$ of 2.7‰; Kumar et al. 2011). Therefore, the $^{15}N$-depleted values of PON during stratification in August indicate greater relative uptake of $NO_3^-$ in the productive surface waters perhaps due to decreases (via microbial utilization or conversion to oxidized form, thus nitrate) in the available ammonium or a shift in the phytoplankton species composition (Kumar et al. 2011). The $^{15}N$-enrichment of PON with depth during stratification is due to the combination of less nitrate utilization in the less productive colder deeper waters, and also the preferential release of $^{14}N$ from remineralization of sinking POM (Kumar et al. 2011).

### 3.4.4 New insights into OC budget of Lake Superior

The current OC budget for Lake Superior is thought to be out of balance with estimated community respiration far exceeding total inputs of OC from primary production and allochthonous sources. Cotner et al. (2004) reported total OC input of 6-9.23 Tg C yr$^{-1}$ (inclusive of atmospheric deposition, riverine loading, and phytoplankton primary production), and total OC output of 13.48-40.5 Tg C $yr^{-1}$, mainly from respiration Urban et al. (2005) reported total OC input of 2.4-7.7 Tg $Cyr^{-1}$ (inclusive of shoreline erosion in addition to the input terms in the earlier estimate), and total OC output of 13.2-83.1 Tg C $yr^{-1}$. The mismatch in the input and output OC terms has been attributed to poor constraint on the respiration and/or OC pool sizes (because of poor spatial and temporal coverage in existing data, mostly nearshore-biased), or could result from the lake's OC budget not being in equilibrium.

Based on our data on the abundances of OC, we determine that the lake's DOC pool is 14.2-16.4 Tg C (compared to its DIC pool of 121-122 Tg C), and its POC pool is 0.9-1.3 Tg C. Our values are slightly lower than, but generally comparable to, the DOC and POC pools of 15-19 Tg C and 1.2-2.5 Tg C estimated by Cotner et al. (2004), and ~ 17 Tg C and ~1 Tg C, respectively, given by Urban et al. (2005). Recent work by Sterner (2010) constrains the total phytoplankton production in the lake at 9.73 Tg C $yr^{-1}$. Using this new primary production and the flux OC flux estimates of Cotner et al. (2004) puts the total OC inputs at 10.43-10.76Tg C $yr^{-1}$. Similarly, recalculating using the other input terms of Urban et al. (2005) with the new primary production and pool estimates gives OC input of 10.13-10.73 Tg C $yr^{-1}$. These values are justly slightly lower than the lower limit of the respiration estimate (13 Tg C $yr^{-1}$) by both Cotner et al. (2004) and Urban et al. (2005) and considerably lower than their upper limits (39-81 Tg C $yr^{-1}$).

In an effort to constrain the respiration estimate in the lake, Strom (2007) used changes in deep water oxygen and nitrate levels at multiple offshore and nearshore sites from 2003-2006 and reported an annual lake-wide community respiration of ~4 Tg C $yr^{-1}$. Also, Bennington (2010) using a modeling approach, reported a modeled lake-wide community respiration of 5.8 Tg C $yr^{-1}$ based on data from the year 1999. These revised respiration estimates imply that poor constraint (thus, over-estimation) on the lake-wide community respiration could have been the cause of the apparent imbalance in the lake's OC budget as reported previously by Cotner et al. (2004) and Urban et al. (2005). We initially hoped that spatial and temporal variations within the radiocarbon in lake DIC could provide further insight into these respiration values. However, due to the large amount of carbon in the DIC pool and the magnitude of atmospheric exchange processes, radiocarbon measurements on DIC do not provide sufficient resolution to further constrain the lake's respiration. It is worth noting that winter time OC abundances, primary production, and community respiration are still unconstrained because portions of the lake show incomplete ice-coverage making sampling more challenging by ice or water. Therefore winter time and higher spatial scale community respiration in the lake would be needed to potentially, definitely close the OC budget of the lake.

### 3.4.5 Putative origins of POC

The depletion of $\Delta^{14}C$ values of POC compared to DOC shows a spatial decoupling of the processes modulating their dynamics and radiocarbon content. The $^{14}C$ ages of POC in the lake (Fig. 3-5E, F) suggest that they are affected by an old C source (considerably pre-bomb) in the water column in June and especially in the deep waters in August. Potential sources of lake POC include autochthonous production, terrestrial organic matter, dry and wet atmospheric deposition, and sedimentary particulate matter. In the paragraphs below, we will discuss these sources in detail to determine their relative importance in Lake Superior.

If the majority of the POC were derived from that year's autochthonous production, it would have $\Delta^{14}C$ values tracking lake $\Delta^{14}C_{DIC}$ since the inorganic carbon is incorporated during biosynthesis (McCarthy et al. 2011). The $^{14}C$-depletion of POC compared to DIC implies that a variable portion of the POC in the lake is not derived from recent within-lake production. Lake Superior is less influenced by organic carbon from terrestrial origin than most lake systems, due in large part to its low watershed-to-lake-surface-area-ratio (Cotner et al. 2004), with current estimates of terrestrial organic carbon inputs at ~10% of total annual organic carbon influxes (Maier and Swan 1978; Urban et al. 2005; Cotner et al. 2004; Zigah et al. 2011). $\Delta^{14}C$ values of POC from most of the nearshore sites were not considerably $^{14}C$-depleted (Fig 3-5E, F) and storm flow data from a Lake Superior tributary (Amity Creek, Duluth, MN) also shows moderate $\Delta^{14}C_{POC}$ ($\Delta^{14}C$=11.1‰; Zigah et al. 2011). Thus both existing carbon budgets and isotopic data discount terrestrial influence as the likely cause of the $^{14}C$ depletion of open lake POC samples.

The atmosphere could be of importance in the Lake Superior carbon budget because of the huge surface area of the lake compared to the watershed (Cotner et al. 2004); however, organic carbon loading from the atmosphere to Lake Superior has not been directly studied. Estimates of atmospheric organic carbon input range from 0.3-4% (Urban et al. 2005) to 6-34% (Cotner et al. 2004) of the lake organic carbon and these estimates include both dissolved and particulate phases. Studies of rainwater at Wilmington, NC, USA have shown that unfiltered rainwater mostly consists of dissolved organic carbon (~98%; Wiley et al. 2000); therefore it is likely that only trace POC is potentially derived from wet deposition. The eastern and central provinces of the lake fall within the downwind range of depositional influence of the Chicago, IL metropolitan area, and previous studies have noted inputs of hydrophobic organic contaminants (HOCs) and soot carbon (both of which are byproducts of industrial incineration processes) to this region of the lake from the Chicago urban area (Buckley et al. 2004). Soot carbon (also, black carbon) from industrial sources could be considerably $^{14}C$-depleted (Reddy et al. 2002) and could shift POC values towards the old $^{14}C$ age ranges seen in our POC samples. However, our data shows that black carbon makes up only 4-9% of POC and was on average, significantly lower than the amount needed to explain the old ages of POC seen in the lake, assuming that all the measured black carbon was $^{14}C$-dead ($\Delta^{14}C$ = -1000‰), i.e, from fossil fuels rather than wood/biomass combustion. This is a conservative assumption considering Lake Superior's location bordering boreal forest, the prevalence of wildfires in the region, and the low human population in the Lake Superior watershed.

The lake's underlying sediments are another, and more likely, source of older particulate material in the lake. Both radioactive decay of organic $^{14}C$ in sediments (not so significant in a relatively young lake such as Lake Superior, approximately 10,000 years old (Mothersill 1988; Lowell et al. 1999) and preferential diagenesis of more labile younger organic carbon can result in $^{14}C$-depleted (old) values of organic carbon in the sediments (McCarthy et al. 2011; Eglinton et al. 1997). Lake Superior receives a considerable amount of eroded materials each year, and previous studies have documented resuspension events in different parts of the lake (Baker and Eisenreich 1989; Hawley 2000; Urban et al. 2004; Zigah et al. 2011). Of the estimated 40 million metric tonnes of eroded materials from the US coastline of the Great Lakes every year, about a quarter (5-11 million metric) goes into Lake Superior (Stortz et al. 1976; Monteith and Sonzogni 1976; Siebel et al. 1976; Kemp et al. 1978). The southwestern shoreline of Lake Superior contains large (~3600 km$^2$) deposits of glacial-lacustrine red clay which is susceptible to extensive erosion into the lake (Siebel et al. 1976; Bahnick et al. 1978). By using Landsat Satellite data and measurements of suspended solids, particle sizes, and turbidity depth profiles, Stortz et al. (1976) reported substantial sedimentary inputs into the western Lake Superior water column, especially during winter storms, and high northeasterly winds. In a study off the Keweenaw Peninsula, Churchill and Williams (2004) used a Benthic Acoustic Stress Sensor

(BASS) array (which had current and pressure sensor) and an in situ particle sampler for collecting suspended particulate matter, both deployed just above the lake floor on a tripod stands, and reported that high bottom stress during autumn and winter results in sediment resuspension (even at 90 m depth), and these resuspended materials tend to be carried offshore by the Keweenaw current which flows northeastward along the western shoreline of Keweenaw Peninsula. The Keweenaw current (a coastal jet) hence inputs these resuspended particulate materials into the eastern basin of Lake Superior.

Similarly, using a submersible equipped with video, camera, and a sampler, Flood (1989) reported large arcuate depressions at 240 m and furrows at 90 m water depth sites in the central basins, consistent with observations from side-scan sonar (Flood and Johnson 1984), and caused by extensive erosion and resuspension of sediments. Also, Hawley (2000) also reported sedimentary resuspension and transport offshore in a study off the Keweenaw Peninsula. Urban et al. (2005) reported that much of the POC in Lake Superior was derived from sediment resuspension based on concentrations of total suspended particles (TSP), and the observed relationship between POC and TSP. Resuspended materials have also been seen in sediment traps collecting sinking POC in the Lake. For instance, between 10-30% of sinking POC just offshore of the Keweenaw Peninsula in Lake Superior has been reported to be derived from resuspended sediments (Urban et al. 2004).

In this study, coupling the considerable $^{14}C$-depletion in POC with depth during stratification in August with the presence of old POC in samples from both surface and deep water during isothermal conditions is consistent with both water-column diagenesis/OC-remineralization of freshly synthesized OC, and considerable sediment resuspension, where such resuspension is likely confined to the hypolimnion by density-driven stratification in August and present throughout the water column during isothermal periods. Our data indicates that resuspension affects open-lake sites as well. That open-water regions can be significantly impacted has been seen in marine systems as well; as much as 35% of sinking POC from several open-ocean sites has been reported to be derived from resuspended sediments (Hwang et al. 2010).

### 3.4.6 Spatial heterogeneity of $\Delta^{14}C$ of POC (POC ages) in Lake Superior

POC ages varied widely across the lake, and with the physical regime of the lake (Fig. 3-7E, F). Although at a coarse spatial resolution, our data show that POC ages appears to increase ($^{14}C$-depleted) in the deepest basins (SM and EM) of the lake (Fig. 3-6). Trends in the POC ages observed in Lake Superior are consistent with the widespread influence of resuspension on sinking POC as has previously been seen in several open oceanic sites with the oldest marine POC observed in the deepest basin (Hwang et al. 2010). In large aquatic systems such as Lake Superior and the oceans, several entrainment, lateral transport and redeposition events are required to deposit sediments in the deeper basins, and that could lead to very old ($^{14}C$-depleted) fine particles being present in these regions. The eastern basin, which exhibits more $^{14}C$ depletion in its POC samples, contains north-south trending troughs covered by up to 13-15 cm of soft sediments sitting on top of stiff post-glacial sediments due to the high currents that flow through (Boyer and Hedrick 1989; Wattrus and Rausch 2001) whereas the western basin, whose POC radiocarbon signatures are more uniform, is characterized by very low relief with the exception of two deep troughs at the northwestern shore, which have thick deposits of sediments spanning several thousand years (Wattrus and Rausch 2001; Pearson 2005).

For the resuspended sediments to considerably influence the lake water as suggested by the old ages of POC, strong bottom currents, and/or strong storms are required. Strong storms occur on the lake during spring and fall when the prevailing winds are out of northeast and are strong enough to erode and entrain sediments (Wattrus and Rausch 2001). Previous studies have shown that the south shore is more susceptible to erosion than the north shore is (Johnson and Johnston 1995) with coastal erosion of clays being about 5 times faster at the south shore (Wattrus and Rausch 2001). Surface water circulation in the lake is mostly cyclonic whereas bottom/deep circulation pattern in the lake is not well-constrained, although thought to be generally cyclonic, flowing northeastward along the south shore (Lam 1978), suggesting that entrained sediments from the south shore would potentially influence the central and eastern regions, those exhibiting older radiocarbon POC values. Although bottom-current velocities have not been well studied in the lake, reported current speeds up to 35 cm $s^{-1}$ at 150 m (mid-depth) and 15 cm $s^{-1}$ at 10 m above the lake floor have been reported near Isle Royale (Carlson 1982) and up to 42 cm $s^{-1}$ off the Keweenaw Peninsula (Miller et al. 1977). The Keweenaw current, a strong alongshelf coastal jet from the Keweenaw Peninsula toward the northeastward direction, periodically splits off toward the center of the lake (Zhu et al. 2001; Van Luven et al. 1999). The velocities of the coastal jet and bottom currents discussed above are high enough to resuspend loose fine sediments in these regions, and to transport them around the central-southern-eastern regions of the lake, leading to the observed considerably aged POC samples at these regions of the lake.

### 3.4.7 Implications of resuspension for ecosystem dynamics and biogeochemistry

The fate of the old POC from resuspended sediments remains an open question. In somewhat smaller lakes (e.g. Lake Michigan), sediment resuspension has been reported to influence water-column ecosystems (Eadie et al. 2002; Schallenberg and Burns 2004). While some studies reported that resuspended sediments increase net primary and winter heterotrophic productions, and community respiration (Cotner et al. 2000; Schallenberg and Burns 2004; Johengen et al. 2008), others noted that the reduced light availability from resuspension events would decrease primary production during the spring season (Lohrenz et al. 2004).

In Lake Superior, the DIC pool (121-122 Tg C) is about two orders of magnitude larger than the POC pool (0.9-1.3 Tg C), so mineralization or sinking of the old POC to DIC will have only trace effects on the concentration and isotopic values of the DIC. For instance, if all the POC at each site and depth is mineralized to DIC, the effect on the $\Delta^{14}C$ of DIC at each site and depth (in both stratified and mixed lake) will be $^{14}C$-depletion by only 0.5 ± 0.4‰ (range: 0.02-2.06‰, n = 38). Although this estimation is conservative as we do not expect all the old POC to be mineralized, the radiocarbon effect, nonetheless, is negligible as these values are below the precision of $\Delta^{14}C$ of DIC of 2-5‰ in this study. Similarly, if all the POC at each site and depth is photochemically and/or microbially degraded to DOC (again, a conservative estimate), the effect on $\Delta^{14}C$ of DOC will be 2.5 ± 3.3‰ (range: 0.04-17.34‰, n = 39). While the POC effect on DOC radiocarbon values would be negligible (≤5‰) in most cases, it would be substantial (~17‰) at the EM site (40 m depth) during mixed lake condition if all the POC was degraded to DOC.

Therefore, in a further attempt to identify whether old POC acts as s a reduced carbon or food subsidy to the Lake Superior ecosystem, $\Delta^{14}C$ values of mesozooplankton from the lake were measured. Mesozooplanton biomass does not exhibit incorporation of the old carbon (Zigah et al. 2011). Therefore, old POC does not appear to be supporting zooplankton secondary production (and by extension, zooplanktivorous fish production) in the lake.

As discussed above, carbon sources fueling respiration in the lake are not well constrained. If the lake's resuspended sediments do fuel part of this respiration, then the lake's response to yearly variations in primary productivity would be buffered. More detailed studies (including investigation of bacterial production and respiration) to constrain the degree of benthic-pelagic coupling in Lake Superior's carbon cycle should be undertaken.

## 3.5 Conclusion

Our lakewide radiocarbon and stable carbon data, the first of its kind in a large-lake system, reveals that bulk DOC across the entire lake (western through eastern basins) had modern radiocarbon signatures and a turnover time of $\leq$ 60 years, indicating most of the DOC in the lake is semi-labile. The semi-lability of DOC is in sharp contrast to the old DOC ages seen in oceanic systems. This may be due in part to different water residence times (~170 years for Lake Superior (Quinn 1992) compared to ~1000-1500 years for the deep ocean (McNichol and Aluwihare 2007), but also results at least in part from the dimictic nature of the lake, which homogenizes the water column twice a year, bringing a portion of DOC from deep water up to the more reactive (both microbially and photochemically) surface waters much more often and at a higher proportion than can occur in the open ocean. Variations in DOC concentrations across the lake support the idea that high resolution data on respiration estimates are needed in order to have a quantitative understanding of the OC budget of the lake.

Our POC data show lateral heterogeneity in the mean ages and sources of POC across Lake Superior. The oldest POC samples were mostly seen in the water column of the deepest basins (eastern and southern) in the lake. POM characteristics such as POC and PON concentrations, $\delta^{15}N$, $\delta^{13}C$, and $\Delta^{14}C$ were tightly coupled to the physical regime of the lake, with $^{14}C$-depletion of POM with depth observed during stratification. The extent of $^{14}C$-depletion of POM appears to be related to the total water depth. Black carbon appears to be only a trace portion of the POC, and cannot account for the old ages of POC seen in the lake. Our results support previous observations (eg., Baker and Eisenreich 1989; Hawley 2000; Urban et al.

2004) that a variable and often substantial portion of POC in the lake, especially in the hypolimnetic waters and in the deeper basins, is derived from sediment resuspension. The ultimate fate of these resuspended materials should be further evaluated by examining the $^{14}C$ signatures within bacterial respiration and/or bacterial biomass, and the biomass of known benthic feeders within the zooplankton and fish communities.

**Table 3-1. Coordinates, sampling, and total water depths of stations in Lake Superior. Numbers in parentheses indicate August water sampling depth if it differed from the June value.**

| Site name | Site description | Latitude (degrees and minutes N) | Longitude (degrees and minutes W) | Depth sampled (m) | Total water depth (m) |
|---|---|---|---|---|---|
| WM | Western mooring/site (Offshore) | 47 ° 18.835 | 89 ° 51.078 | 5, 30, 127 | 171 |
| CM | Central mooring (Offshore) | 48 ° 01.916 | 87 ° 44.409 | 5, 30 (35), 190 | 257 |
| EM | Eastern mooring (Offshore) | 47 ° 33.694 | 86 ° 38.739 | 5, 40, 210 | 242 |
| SM | Southern mooring (Offshore) | 46 ° 54.410 | 86 ° 36.017 | 5, 40, 340 | 398 |
| NM | Northern mooring (Offshore) | 48 ° 29.351 | 87 ° 3.802 | 5, 30, 150 | 216 |
| BR | Baptism River mouth (Nearshore) | 47 ° 20.056 | 91 ° 11.423 | 4 (2), 14 | 20 |
| ONT | Ontonagon River mouth (Nearshore) | 46 ° 53.705 | 89 ° 20.401 | 4 (5), 13 | 20 |
| NB | Nipigon Bay (Nearshore) | 48 ° 51.646 | 87 ° 45.344 | 4 (5), 10, 50 | 62 |

**Table 3-2. Concentrations and isotopic compositions of DIC, DOC, and POM from multiple open lake and nearshore sites in the isothermal (well-mixed) water column of Lake Superior in June 2009. Precision of radiocarbon are based on analyses of multiple external standards, and those of concentrations are based on multiple sample analyses; 'nd' means not determined either because sample was not collected, not measured or lost during processing in the field or lab; OL refer open lake samples; and NS means nearshore samples.**

| Station | Depth | DIC | | | DOC | | | | POC | | | | PON | | POM |
|---|---|---|---|---|---|---|---|---|---|---|---|---|---|---|---|
| | (m) | µM | $\delta^{13}C$ (‰) | $\Delta^{14}C$ (‰) | µM | $\delta^{13}C$ (‰) | Measured $\Delta^{14}C$ (‰) | Blank-corrected $\Delta^{14}C$ (‰) | µM | $\delta^{13}C$ (‰) | Measured $\Delta^{14}C$ (‰) | Blank-corrected $\Delta^{14}C$ (‰) | µM | $\delta^{15}N$ (‰) | molar C:N |
| June 2009 - Isothermal | | | | | | | | | | | | | | | |
| | | | | | | | | | | | | | | | |
| WM - OL | 5 | 825 ± 3 | 0.4 | 56 ± 4 | 94.0 ± 1.3 | -26.5 | 49 ± 4 | 93 ± 4 | 4.77 | -29.2 | 21 ± 4 | 50 ± 5 | 0.66 | 1.7 | 7.2 ± 0.7 |
| CM - OL | 5 | 830 ± 2 | 0.3 | 61 ± 4 | 90.8 ± 0.8 | -25.9 | 58 ± 4 | 104 ± 4 | 4.67 | -29.9 | -55 ± 4 | -45 ± 5 | 0.59 | 0.1 | 7.9 ± 0.8 |
| EM - OL | 5 | 819 ± 2 | 0.4 | 59 ± 5 | 90.8 ± 0.7 | -26.3 | 42 ± 4 | 86 ± 4 | 4.44 | -30.3 | -24 ± 3 | -4 ± 4 | 0.56 | 0.1 | 7.9 ± 0.7 |
| SM - OL | 5 | 824 ± 1 | 0.4 | 62 ± 3 | 90.8 ± 0.6 | -26 | 25 ± 5 | 68 ± 5 | 4.56 | -29.8 | 34 ± 5 | 68 ± 6 | 0.57 | 0.4 | 8.0 ± 0.8 |
| NM - OL | 5 | 826 ± 2 | 0.5 | 52 ± 2 | 94.2 ± 0.6 | -26.5 | 22 ± 4 | 63 ± 4 | 5.13 | -29.9 | 34 ± 3 | 64 ± 4 | 0.74 | 0.7 | 7.0 ± 0.6 |
| ONT - NS | 4 | 851 ± 3 | nd | 38 ± 2 | 129.2 ± 8.2 | nd | -10 ± 3 | 17 ± 3 | 14.47 | -29.3 | 16 ± 3 | 24 ± 3 | 1.57 | -0.6 | 9.1 ± 0.8 |
| BR - NS | 4 | 822 ± 2 | nd | 54 ± 4 | 110.8 ± 0.9 | nd | 38 ± 4 | 74 ± 4 | 6.85 | -30.6 | 14 ± 4 | 32 ± 5 | 0.72 | 1.1 | 9.5 ± 0.9 |
| | | | | | | | | | | | | | | | |
| WM - OL | 30 | 830 ± 2 | 0.3 | 60 ± 4 | 93.3 ± 1.3 | -26.4 | 40 ± 5 | 83 ± 5 | 4.81 | -29.4 | -12 ± 4 | 8 ± 5 | 0.57 | 1.2 | 8.4 ± 0.8 |
| CM - OL | 30 | 824 ± 3 | 0.4 | 64 ± 4 | 91.7 ± 0.8 | -26.3 | 36 ± 4 | 79 ± 4 | 4.55 | -29.6 | 25 ± 3 | 56 ± 4 | 0.64 | 1 | 7.1 ± 0.7 |

| | | | | | | | | | | | | | | | |
|---|---|---|---|---|---|---|---|---|---|---|---|---|---|---|---|
| EM - OL | 40 | 827 ± 1 | 0.3 | 54 ± 4 | 90.0 ± 1.4 | -26.3 | 39 ± 4 | 84 ± 4 | 4.80 | -29.8 | -303 ± 3 | -354 ± 4 | 0.61 | 0.3 | 7.8 ± 0.8 |
| SM - OL | 30 | 825 ± 1 | 0.4 | 56 ± 3 | 88.3 ± 1.1 | -26.4 | 40 ± 4 | 86 ± 4 | 4.20 | nd | -16 ± 4 | 8 ± 5 | 0.55 | -1 | 7.7 ± 0.6 |
| NM - OL | 30 | 827 ± 1 | 0.5 | 61 ± 4 | 94.2 ± 0.5 | -26.3 | 58 ± 4 | 102 ± 4 | 5.09 | -29.4 | 10 ± 4 | 34 ± 5 | 0.70 | 0 | 7.3 ± 0.7 |
| | | | | | | | | | | | | | | | |
| WM - OL | 127 | 822 ± 2 | 0.3 | 58 ± 3 | 93.3 ± 1.2 | -26.7 | 42 ± 4 | 85 ± 4 | 4.67 | -29.9 | 12 ± 4 | 39 ± 5 | 0.62 | 1.6 | 7.5 ± 0.7 |
| CM - OL | 190 | 830 ± 4 | 0.4 | 63 ± 4 | 91.7 ± 0.5 | -26.6 | 35 ± 4 | 78 ± 4 | 4.28 | -29.1 | -107 ± 4 | -112 ± 5 | 0.56 | 1.2 | 7.6 ± 0.7 |
| EM - OL | 210 | 823 ± 3 | 0.3 | 58 ± 2 | 89.2 ± 0.7 | -26 | 30 ± 4 | 74 ± 4 | 4.02 | -29 | -172 ± 3 | -196 ± 4 | 0.48 | 3.7 | 8.4 ± 0.7 |
| SM - OL | 340 | 824 ± 1 | 0.4 | 56 ± 4 | 98.3 ± 0.1 | -26.6 | 52 ± 3 | 93 ± 3 | 4.42 | -30 | 1 ± 5 | 27 ± 6 | 0.59 | 0.4 | 7.5 ± 0.7 |
| NM - OL | 150 | 827 ± 0 | 0.5 | 60 ± 4 | 94.2 ± 0.6 | -26.5 | 47 ± 3 | 90 ± 3 | 5.07 | -29.6 | 12 ± 3 | 36 ± 4 | 0.65 | 0.4 | 7.9 ± 0.7 |
| ONT - NS | 13 | nd | nd | nd | 103.3 ± 0.2 | nd | 50 ± 4 | 89 ± 4 | 16.21 | nd | 36 ± 4 | 44 ± 4 | 1.75 | -1.7 | 9.2 ± 0.8 |

**Table 3-3. Concentrations and isotopic compositions of DIC, DOC, and POM from multiple open lake and nearshore sites in the thermally stratified water column of Lake Superior in August 2009. Precision of radiocarbon are based on analyses of multiple external standards, and those of concentrations are based on multiple sample analyses. 'nd' means not determined either because sample was not collected, not measured or lost during processing in the field or lab. OL refer open lake samples, and NS means nearshore samples. DCM is depth of deep chlorophyll maximum during summer thermal stratification.**

| Station | Depth | DIC | | | DOC | | | | POC | | | | PON | | POM |
|---|---|---|---|---|---|---|---|---|---|---|---|---|---|---|---|
| | (m) | µM | $\delta^{13}C$ (‰) | $\Delta^{14}C$ (‰) | µM | $\delta^{13}C$ (‰) | Measured $\Delta^{14}C$ (‰) | Blank-corrected $\Delta^{14}C$ (‰) | µM | $\delta^{13}C$ (‰) | Measured $\Delta^{14}C$ (‰) | Blank-corrected $\Delta^{14}C$ (‰) | µM | $\delta^{15}N$ (‰) | molar C:N |
| August 2009 - Stratified | | | | | | | | | | | | | | | |
| Epilimnion | | | | | | | | | | | | | | | |
| WM - OL | 5 | 808 ± 0 | 1.3 | 61 ± 3 | 104.0 ± 2.6 | -26.1 | 51 ± 3 | 90 ± 3 | 11.96 | -27.8 | 33 ± 4 | 44 ± 4 | 1.42 | -2.8 | 8.4 ± 0.7 |
| CM - OL | 5 | 815 ± 2 | 1 | 62 ± 3 | 121.6 ± 2.8 | -26 | 50 ± 4 | 82 ± 4 | 11.61 | -29.1 | 39 ± 3 | 51 ± 3 | 1.40 | -2.3 | 8.3 ± 0.7 |
| EM - OL | 5 | 810 ± 2 | 0.9 | 59 ± 4 | 88.8 ± 1.0 | -26 | 54 ± 3 | 101 ± 3 | 9.09 | -29.2 | 38 ± 4 | 53 ± 4 | 1.20 | -3.9 | 7.6 ± 0.6 |
| SM - OL | 5 | 803 ± 1 | 1.1 | 54 ± 4 | 116.0 ± 1.8 | -26 | 27 ± 4 | 60 ± 4 | 8.59 | -28.1 | -24 ± 3 | -15 ± 3 | 1.02 | -2.9 | 8.4 ± 0.7 |
| NM - OL | 5 | 812 ± 2 | 0.9 | 50 ± 3 | 120.7 ± 2.2 | -26.1 | 21 ± 3 | 52 ± 3 | 12.39 | -28.9 | 22 ± 3 | 31 ± 3 | 1.46 | -2.9 | 8.5 ± 0.7 |
| ONT - NS | 5 | 810 ± 2 | 1 | 56 ± 4 | 124.7 ± 2.1 | -28.3 | nd | nd | 10.33 | -27.2 | nd | nd | 1.25 | -3.7 | 8.3 ± 0.7 |
| BR - NS | 2 | 789 ± 3 | 0.9 | 60 ± 4 | 208.0 ± 2.7 | -26.5 | 74 ± 4 | 94 ± 4 | 13.40 | -28.3 | 7 ± 3 | 15 ± 3 | 1.60 | -2.7 | 8.4 ± 0.7 |
| NB - NS | 5 | 871 ± 3 | 0.2 | 36 ± 3 | 128.1 ± 2.7 | -26.5 | 39 ± 4 | 69 ± 4 | 16.50 | -28 | -19 ± 4 | -14 ± 4 | 2.09 | -2 | 7.9 ± 0.7 |
| DCM* | | | | | | | | | | | | | | | |
| WM - OL | 30 | 826 ± 2 | 0.6 | 59 ± 3 | 102.6 ± 2.6 | -26 | 46 ± 3 | 85 ± 3 | 9.61 | -29.7 | 30 ± 3 | 44 ± 3 | 1.25 | -1.4 | 7.7 ± 0.6 |
| CM - OL | 35 | 830 ± 3 | 0.4 | 59 ± 5 | 86.5 ± 1.6 | -26.1 | 50 ± 3 | 97 ± 3 | 7.23 | -30.5 | 40 ± 3 | 61 ± 3 | 0.88 | -1.4 | 8.2 ± 0.7 |
| EM - OL | 40 | 823 ± 1 | 0.5 | 60 ± 3 | 96.3 ± 1.8 | -26.3 | 44 ± 4 | 86 ± 4 | 9.82 | -34.9 | 58 ± 11 | 75 ± 12 | 1.18 | -1.6 | 8.3 ± 0.7 |

| | | | | | | | | | | | | | | | |
|---|---|---|---|---|---|---|---|---|---|---|---|---|---|---|---|
| SM - OL | 30 | 813 ± 3 | 0.6 | 65 ± 4 | 85.9 ± 1.2 | -26 | 41 ± 3 | 88 ± 3 | 11.76 | -31 | 29 ± 3 | 40 ± 3 | 1.41 | -1.7 | 8.3 ± 0.7 |
| NM - OL | 30 | 816 ± 3 | 0.6 | 56 ± 4 | 107.8 ± 1.5 | -25.9 | 40 ± 5 | 76 ± 5 | 11.21 | -29.5 | 16 ± 4 | 26 ± 4 | 1.30 | -1.6 | 8.6 ± 0.7 |
| NB - NS | 10 | 871 ± 1 | nd | nd | 129.3 ± 1.5 | -26 | 52 ± 4 | 83 ± 4 | 10.20 | -29 | 3 ± 3 | 13 ± 3 | 1.29 | -1.5 | 7.9 ± 0.7 |
| Hypolimnion | | | | | | | | | | | | | | | |
| WM - OL | 127 | 831 ± 2 | 0.2 | 52 ± 5 | 89.3 ± 1.3 | -25 | 14 ± 4 | 56 ± 4 | 2.93 | -29.5 | -17 ± 3 | 20 ± 4 | 0.34 | 3.1 | 8.7 ± 0.7 |
| CM - OL | 190 | 830 ± 1 | 0.2 | 55 ± 4 | 112.1 ± 0.9 | -23.7 | 24 ± 4 | 58 ± 4 | 2.29 | -29 | -29 ± 3 | 18 ± 5 | 0.25 | 0.9 | 9.2 ± 0.8 |
| EM - OL | 210 | 824 ± 2 | 0.3 | 63 ± 4 | 109.5 ± 1.7 | -25.9 | 46 ± 4 | 82 ± 4 | 3.16 | -29.3 | -20 ± 4 | 13 ± 6 | 0.37 | -1.4 | 8.6 ± 0.7 |
| SM - OL | 340 | 827 ± 2 | 0.1 | 61 ± 4 | 85.7 ± 1.4 | -26.1 | 44 ± 3 | 92 ± 3 | 2.88 | -32.3 | -145 ± 11 | -170 ± 16 | 0.32 | -0.1 | 8.9 ± 0.7 |
| NM - OL | 150 | 824 ± 2 | 0.2 | 65 ± 4 | 83.1 ± 1.2 | -26.1 | 44 ± 4 | 94 ± 4 | 2.58 | -28.6 | -0 ± 4 | 55 ± 6 | 0.33 | 0.9 | 7.8 ± 0.7 |
| BR - NS | 14 | 805 ± 4 | 1 | 57 ± 3 | 98.4 ± 1.7 | -26 | 39 ± 3 | 79 ± 3 | 13.38 | -28 | 28 ± 3 | 37 ± 3 | 1.60 | -2.1 | 8.4 ± 0.7 |
| NB - NS | 50 | 841 ± 2 | -0.1 | 54 ± 3 | 135.3 ± 2.6 | -26 | 39 ± 3 | 68 ± 3 | 5.94 | -29.8 | 16 ± 4 | 37 ± 5 | 0.76 | -2.8 | 7.8 ± 0.7 |

**Table 3-4. Black carbon (BC) content of suspended POC from June (isothermal condition) sampling in the water column of Lake Superior. BC was measured via thermal oxidation at 375 °C for 24 hrs.**

| Station | Depth (m) | BC (μM) | POC (μM) | BC/POC (mole %) |
|---|---|---|---|---|
| EM - OL | 5 | 0.17 | 4.44 | 3.9 |
| EM - OL | 40 | 0.43 | 4.80 | 8.9 |
| EM - OL | 210 | 0.23 | 4.02 | 5.6 |
| CM - OL | 5 | 0.21 | 4.67 | 4.5 |
| CM - OL | 190 | 0.18 | 4.28 | 4.1 |
| SM - OL | 30 | 0.20 | 4.20 | 4.8 |
| WM - OL | 30 | 0.20 | 4.81 | 4.1 |

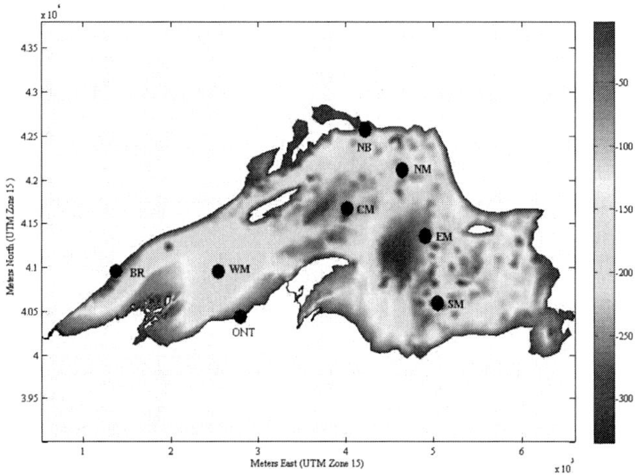


Fig. 3-1. Bathymetry map of Lake Superior showing stations in the open lake, and nearshore sites. The open lake sites include western station/mooring (WM), central mooring (CM), northern mooring (NM), eastern mooring (EM), and southern mooring (SM). The nearshore sites are off Baptism River (BR), off Ontonagon River (ONT), and off Nipigon Bay (NB). The contours are water depth in meters.

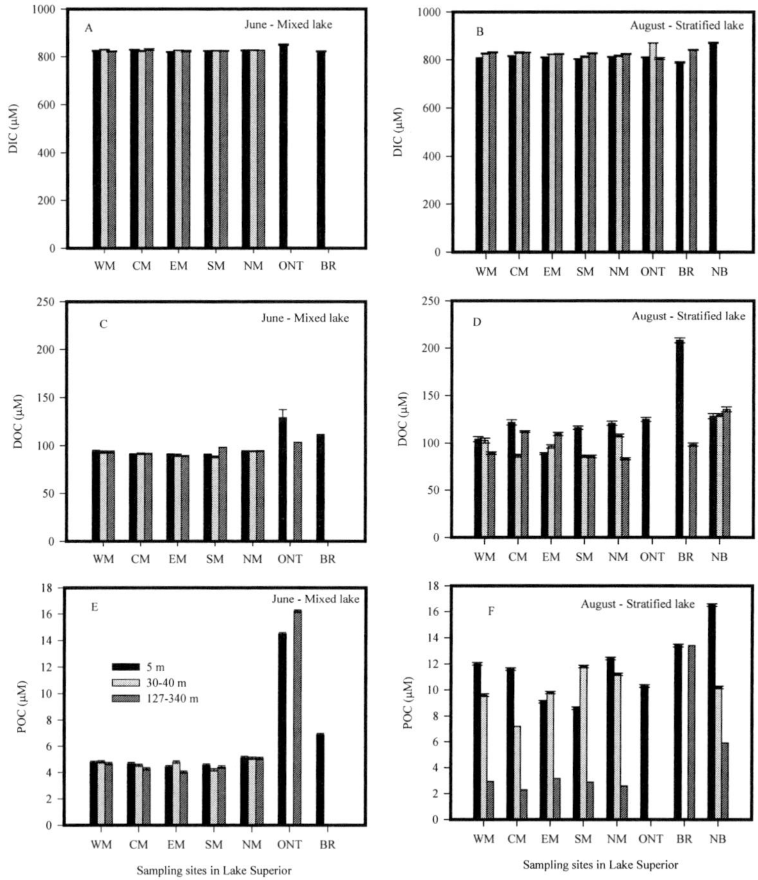


Fig. 3-2. Site-specific distributions of the isothermal lake concentrations of (A) DIC, (C) DOC, (E) POC, and stratified lake concentrations of (B) DIC, (D) DOC, (F) POC. Isothermal samples were collected during June 2009, and stratified samples taken during August 2009. The depth labels in the plot are for offshore samples. The equivalent nearshore depths are 2-4 m for the surface, 10 m at NB sites-the only nearshore site with mid-depth value, and 13-50 m for deep samples

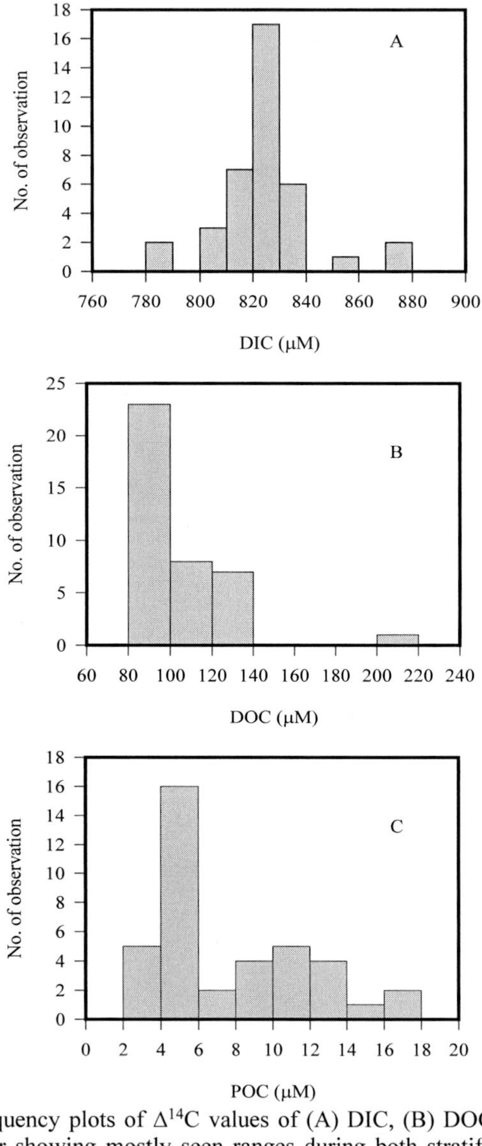


Fig. 3-3. Frequency plots of $\Delta^{14}C$ values of (A) DIC, (B) DOC, and (C) POC in Lake Superior showing mostly seen ranges during both stratified and isothermal seasons. Note that $\Delta^{14}C$ of DOC and POC are the measured values.

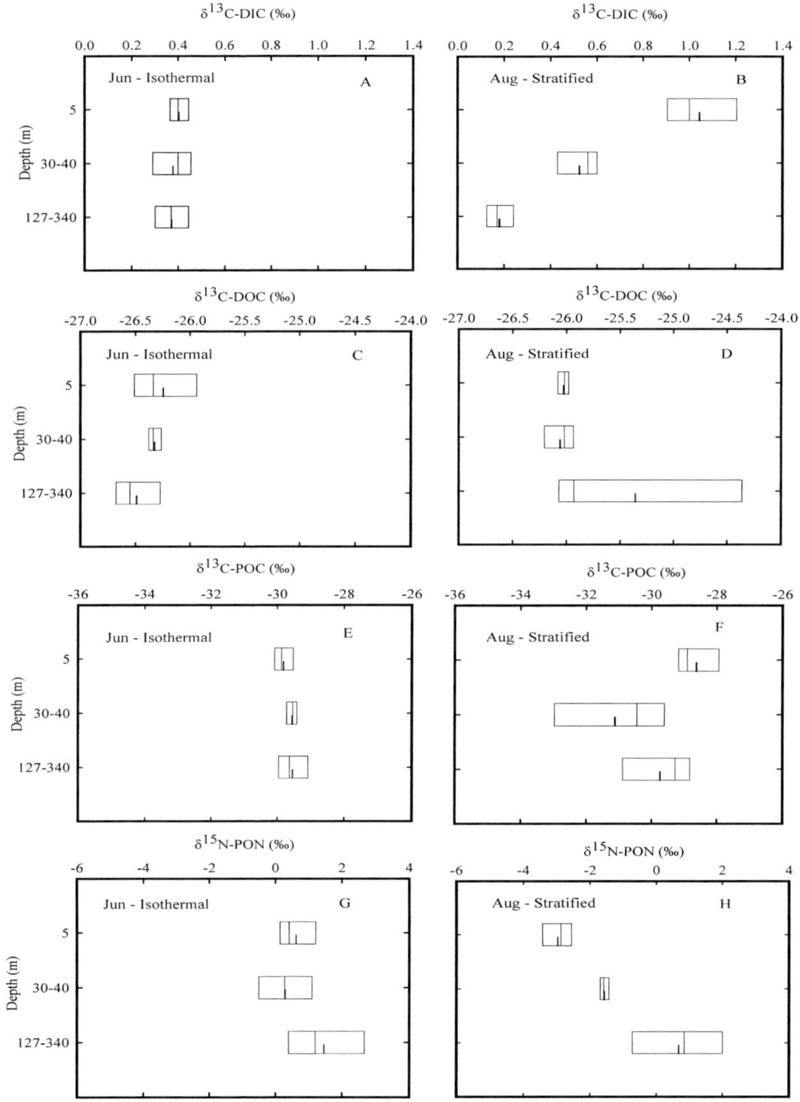


Fig. 3-4. Depth profiles of mean values of all sites during isothermal lake condition in June of (A) $\delta^{13}C_{DIC}$ (C) $\delta^{13}C_{DOC}$ (E) $\delta^{13}C_{POC}$, and (G) $\delta^{15}N_{PON}$ and during stratified lake condition in August of (B) $\delta^{13}C_{DIC}$ (D) $\delta^{13}C_{DOC}$ (F) $\delta^{13}C_{POC}$, and (H) $\delta^{15}N_{PON}$. Box represents the 25th and 75th percentiles, the solid line within the box represents the median value, and the dash line within the box is mean value.

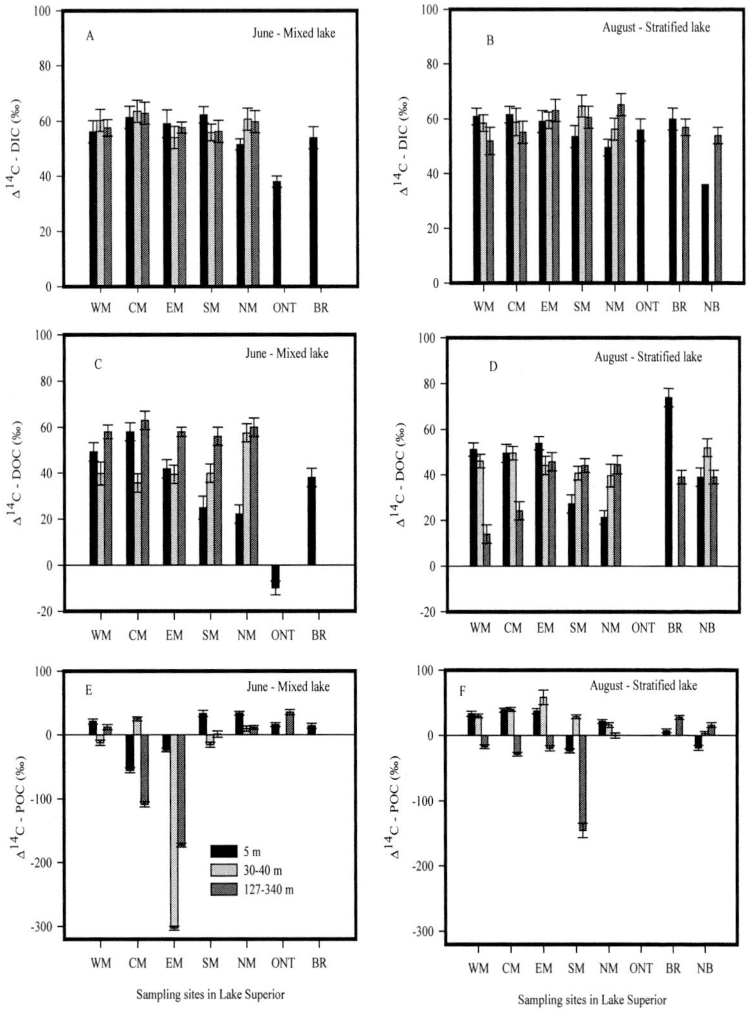


Fig. 3-5. Site-specific distribution of the isothermal lake measured $\Delta^{14}C$ values of (A) DIC, (C) DOC, (E) POC, and stratified lake $\Delta^{14}C$ values of (B) DIC, (D) DOC, (F) POC. The depth labels in the plot are for offshore samples. The equivalent nearshore depths are 2-4 m for the surface, 10 m at NB sites-the only nearshore site with mid-depth value, and 13-50 m for deep samples.

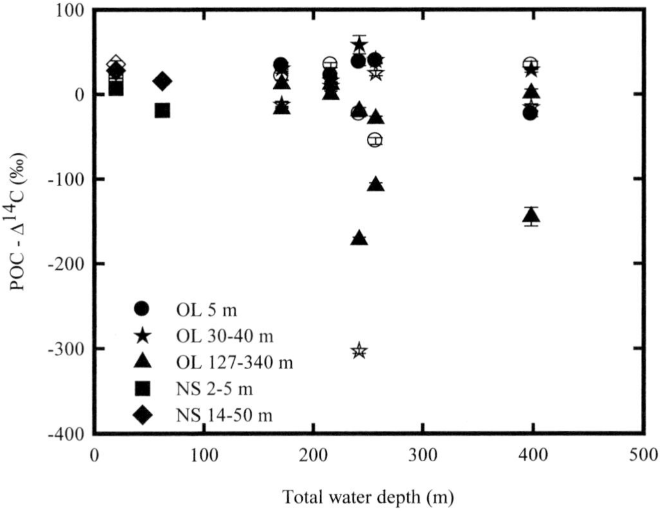


Fig 3-6. Distributions of $\Delta^{14}$C-POC in the water column as a function of total water depth at the various stations. Old POC samples and variability of POC ages appear to increase in the deep basins of the lake. Filled symbols are stratified lake data and open symbols are for isothermal data.

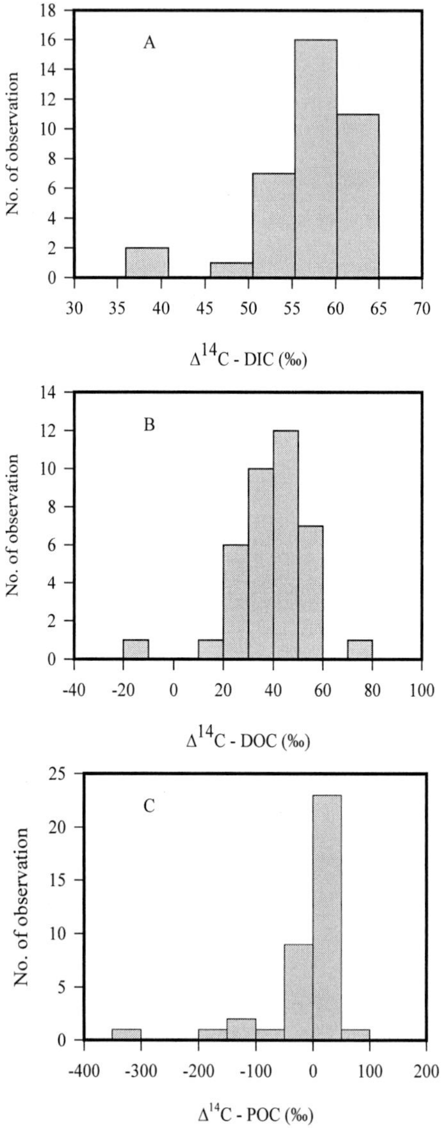


Fig. 3-7. Frequency plots of $\Delta^{14}C$ values of (A) DIC, (B) DOC, and (C) POC in Lake Superior showing mostly seen ranges during both stratified and isothermal seasons. Note that $\Delta^{14}C$ of DOC and POC are the measured values.

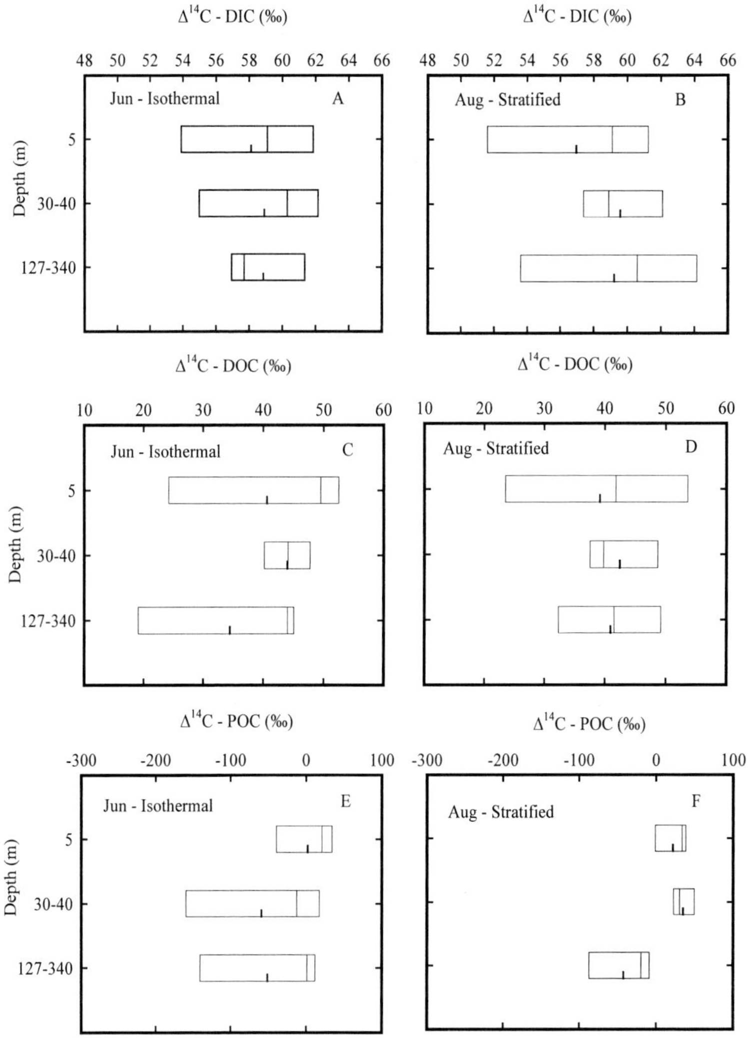


Fig. 3-8. Depth profiles of site-combined-mean values during isothermal condition of (A) $\Delta^{14}C_{DIC}$, (C) $\Delta^{14}C_{DOC}$, (E) $\Delta^{14}C_{POC}$, and during stratified condition of (B) $\Delta^{14}C_{DIC}$ (D) $\Delta^{14}C_{DOC}$ (F) $\Delta^{14}C_{POC}$. The $\Delta^{14}C$ are the measured values. Box represents the 25th and 75th percentiles, the solid line within the box represents the median value, and the dash line within the box is the mean value.

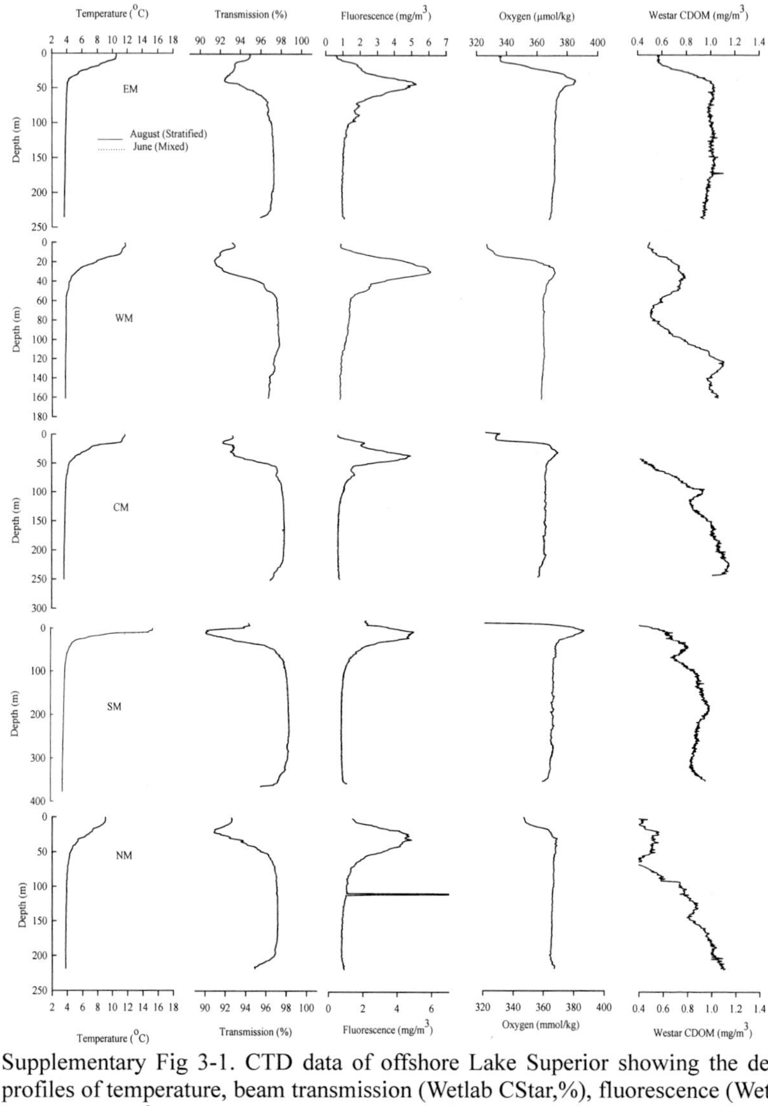


Supplementary Fig 3-1. CTD data of offshore Lake Superior showing the depth profiles of temperature, beam transmission (Wetlab CStar,%), fluorescence (Wetlab Wetstar, mg/m$^3$), dissolved oxygen, and wet CDOM (Wetlab CDOM, mg/m$^3$) in the mixed water column in June and stratified condition in August

# CHAPTER FOUR

# AN INVESTIGATION OF SIZE-FRACTIONATED ORGANIC MATTER FROM LAKE SUPERIOR AND A TRIBUTARY STREAM USING RADIOCARBON, STABLE ISOTOPES AND NMR

## 4.1 Introduction

Organic matter (OM) is a heterogeneous major reactive reservoir in aquatic systems and is ultimately derived from in situ primary production and/or terrigenous sources. OM has numerous sources and biogeochemical sinks in aquatic environments; as a result constraining these is usually difficult, albeit fundamental to our understanding of the global carbon cycle. For instance the quantity of dissolved organic carbon (DOC) in the world's oceans (~685 Gt C) is similar to atmospheric $CO_2$ (Farrington 1992; Hedges 1992; Hansell and Carlson 1998), but a major fraction of oceanic DOC (46-87%) is still uncharacterized at the molecular level (Buffle 1988). Accordingly, the fate of OM, whether it is permanently sequestered in the sediment or remineralized to $CO_2$ is critical to understanding the timescales of the atmospheric $CO_2$ sink with potential implications for regional and global carbon cycles, and climate change through heat exchange and effects on radiative forcing.

It has been widely acknowledged that OM in aquatic systems exists as a continuum of size fractions including particulate, dissolved, colloidal or high molecular weight (HMW), and low molecular weight (LMW) fractions (Maurer 1971; Sharp 1973; Amon and Benner 1996; Guo et al. 1996; Mannino and Harvey 2000; Loh et al. 2004). The size-class partitioning of OM is operationally defined based on filtration and/or ultrafiltration through membranes or cartridges of known nominal pore sizes (Ogura 1970; Ogura 1974; Sigleo et al. 1982; Carlson et al. 1985; Guo et al. 1996; Mannino and Harvey 2000; Minor et al. 2002). Each of the OM size classes may have unique source, sink, chemical and molecular composition, reactivity, and biogeochemical functions in aquatic systems (Skoog and Benner 1997; Guo et al. 1995; Mannino and Harvey 2000; Loh et al. 2006; Guo and Mcdonald 2006). In terms of biogeochemical functions for instance, biodegradation of organic contaminants in wastewater is reported to increase with decreasing size from dissolved, colloidal and particulate fractions (Levine et al. 1985; Chiou et al. 1986). Also, colloidal OM is important in nutrient cycling, pH buffering, and the fate of trace metals and organic pollutants (Sigleo and Means 1990; Dai et al. 1995). Hedges et al. (1994) studied the amino acids and carbohydrate composition in different OM size classes including coarse particulate, fine particulate, and colloidal OM in the Amazon River and noted that these size classes had unique chemical properties.

Although it is widely recognized that molecular size or weight influences the fate and/or microbial utilization or mineralization of OM, this size-lability relationship is not well constrained. While some studies (Wright and Hobbie 1965; Barber 1968; Ogura 1975, 1977; Saunders 1976; Munster and Chrost 1990) have reported that LMW DOM is more bio-reactive and preferentially used by microbes, others have reported that bio-reactivity decreases with size, and that HMW DOM is rather more bio-reactive and is rapidly degraded by microbes (Amon and Benner 1994; Skoog and Benner 1997). For example, Amon and Benner (1996) incubated bacteria with different size fractions of OM from various freshwater and marine environments, and reported greater bacteria growth, respiration, and utilization of HMW DOM compared to LMW DOM, implying a size-reactivity continuum model which states that bio-reactivity increases with increasing OM molecular size. However, bacterial growth efficiency was consistently higher in the LMW DOM fraction and this was attributed to a relatively richer content of organic N and bioavailable elements necessary for bacterial growth (Amon and Benner 1996). Several studies have confirmed the size-reactivity relationship of OM in aquatic systems (Benner et al. 1992; Mannino and Harvey 2000; Hama et al. 2004; Loh et al. 2004). In contrast (or perhaps in refinement) to this size-reactivity model, however, others have reported temporal variability in the relative utilization of HMW vs. LMW DOM by microbes such that a portion of LMW DOM is utilized first, but the proportion of HMW DOM that is mineralized rapidly increases over time (Agura 1977) consistent with other observations that only a portion of HMW DOM is rapidly mineralized, and by extension, some portion persists over longer time (Meyer et al. 1987; Tranvik 1990; Arnosti et al. 1994).

Investigation of the natural abundance radiocarbon ($\Delta^{14}C$) ages and the chemical composition of particulate (both sinking and suspended POM),

HMW and LMW DOM can provide further insight into the roles, sources, bio-reactivity, and fates of different OM size classes. Loh et al. (2004) reported radiocarbon results from the open ocean showing that $^{14}C$ age increases consistently with OM size; the age of OM increased in the sequence from sinking POM (youngest, rapidly cycling) > suspended POM > HMW DOM > LMW DOM (the oldest, cycling slowly). In their study in the Mid-Atlantic Bight (MAB), Aluwihare et al. (2002) noted that in the surface and deep waters, the $\Delta^{14}C$ of HMW DOM was more $^{14}C$-enriched than the bulk DOM by 22‰ and 150‰, respectively. HMW DOM hence represents a modern labile component with younger $^{14}C$ ages. Similar trends in $\Delta^{14}C$ of size-fractionated OM have been seen in other oceanic sites (Guo et al. 1996; Guo et al. 2009; McNichol and Aluwihare 2007).

In contrast to the observations in oceanic systems, several riverine systems do not conform to the OM size-reactivity sequence seen in the oceans. In the Amazon, Hudson, York, and Parker rivers in the US, POC was mostly pre-aged and consistently more $^{14}C$-depleted than concurrent DOC (Raymond and Bauer 2001). Also, the $^{14}C$ age of POC seen in some Northeast US rivers such as Susquehanna, Delaware, and Parker Rivers was much older than co-occurring DOC (Raymond et al. 2004). A similar trend has been observed in other river such as the Hudson River in the US (Caraco et al. 2010) and Strickland and Fly Rivers in Papua New Guinea (Alin et al. 2008).

Mannino and Harvey (2000) examined the biochemical composition (amino acids and carbohydrates) of POM, HMW DOM, and very HMW DOM (VHMW DOM) in the Delaware estuary and noted that the POM consistently had higher polysaccharide and amino acid concentrations than either HMW DOM or VHMW DOM. In a similar study in the Pacific and Atlantic Oceans, Kaiser and Benner (2009), based on amino acids, neutral sugars, and amino sugars compositions of POM, HMW DOM, and LMW DOM reported higher biochemical composition in the POM relative to the DOM fractions. These studies, therefore, extended the size-age and size-reactivity relationships to include a size-composition relationship; the yields or composition of amino acids, neutral sugars, and amino sugars decrease rapidly with decreasing molecular size. Altogether, these three OM size related models incorporate the size, isotopic signature, and chemical composition of OM size fractions such that larger-sized OM is, in general, more bio-reactive, younger in $^{14}C$ age, and has higher yields of these biochemicals in accordance with their relatively unaltered/less altered diagenetic state.

In oceans, large rivers, and estuarine systems, ultrafiltration has been used to partition OM into HMW DOM (usually > 1000 Da) and LMW DOM (usually <1000 Da) to afford examination of the size, chemical compositions, and isotopic signatures in order to constrain their sources, bio-reactivity, fates, and turnover times (Hedges et al. 1994; Guo et al. 1995; Amon and Benner 1996; McCarthy et al. 1996; Benner and Opsahl 2001; Wang et al. 2004; Repeta et al. 2002). Mannino and Harvey (2000) in their study in the Delaware estuary used ultrafiltration to fractionate DOM into high molecular weight (1-30 kDa) and very high molecular weight (30k Da to 0.2 μm) fractions and compared their biochemical compositions to those of concurrent POM in the estuary. In a study in the Mid-Atlantic Bight, Guo et al. (1996) employed ultrafiltration technique to isolate colloidal fractions of different molecular weights (1 kDa to 0.2 μm vs. 10 kDa to 0.2 μm) for further isotopic characterization to determine the cycling of HMW DOM at this site. These studies show that the proportion of DOM that is concentrated in the HMW DOM is 20-30% in the oceans, 13-48% in estuaries, and 24-87% in riverine aquatic systems. In larger, more lentic, more autochthonous aquatic systems such as the oceans, HMW DOM is mostly rich in carbohydrates and nucleic acids, and recycles rapidly because of its recent photosynthetic origin, whereas in lotic or smaller, more terrestrially-influenced lentic systems, humic materials make up a considerable portion of HMW DOM, reducing its bio-reactivity and turn-over times (Amon and Benner 1994; Bianchi et al. 1995; Gustafsson and Gschwend 1997).

Spectroscopic methods such as $^{1}H$ and $^{13}C$ nuclear magnetic resonance (NMR) provide a non-destructive means of studying the chemical compositions of OM in aquatic systems. $^{1}H$ and $^{13}C$ NMR spectra of HMW DOM have proven very useful for determining bulk chemical compositions in terms of major functional groups, and this information has been used to infer bio-reactivity and sources of HMW DOM in the oceans (Benner et al. 1992; McCarthy et al. 1993; Aluwihare et al. 2002; Repeta et al. 2002; Koprivnjak et al. 2009; Gogou and Repeta 2010), estuaries (Abdulla et al. 2010), rivers and streams (Engelhaupt and Bianchi 2001; Repeta et al. 2002; Bianchi et al. 2004), and lakes and reservoirs (Repeta et al. 2002; Wershaw et al. 2005).

Inland aquatic systems affect regional and global carbon cycles through carbon sequestration in sediments and $CO_2$ evasion. Estimated potential global $CO_2$ efflux of 140 Tg C $y^{-1}$ from lakes is roughly half the annual carbon transport from rivers to the ocean (Cole et al. 1994). The importance of lakes as global carbon sequestration 'hotspots' is illustrated by the burial of ~25-58% of carbon that oceans do per year even though the combined surface area of lakes is less than 2% that of the ocean (Dean and Gorham 1998; Cole et al. 2007). In spite of the key role of large lakes in the global carbon cycle, carbon cycling in large lakes of the world, including the Laurentian Great Lakes of North America, remains poorly understood. Large lakes provide a useful comparison to marine and small-lake systems for addressing hypothesized roles of water-body size, oligotrophy, and water-column mixing in mediating aquatic carbon sources and fates. In this study we used the abundances, elemental compositions, and isotopic ($\Delta^{14}C$, $\delta^{13}C$, $\delta^{15}N$) signatures of OM size classes to examine the sources, reactivity, and turnover times of particulate, dissolved, colloidal, and low molecular weight OM in the water column of Lake Superior, the world's largest freshwater lake by surface area. We also used $^{13}C$ NMR and $^{1}H$ NMR spectra of colloidal OM in the lake to assess its compound class composition and potential origin. Differentiating the carbon sources and their reactivity in the lake is critical to understanding the lake's carbon biogeochemistry, and also informs paleoenvironmental studies based on aquatic algal-derived biomarkers and/or $^{14}C$ dating of bulk or organic fractions of sedimentary organic matter.

## 4.2. Materials and methods

### 4.2.1 Study sites

Lake Superior is the Earth's largest freshwater lake by surface area with maximum and mean depths of, respectively, 406 m and 150 m (Urban et al. 2005). It is the coldest and deepest of the Laurentian Great Lakes of North America. The topography of the eastern province of Lake Superior is characterized by north-south trending troughs covered by ~13-15 cm of soft silts and mud overlying stiff post-glacial clays whereas the western province is marked by very low relief (although two deep troughs have been spotted at the northwestern shore) (Boyer and Hedrick 1989; Wattrus and Rausch 2001). The offshore region of the lake is biogeochemically similar to open-ocean locations due to its oligotrophic nature, low terrestrial nutrient loading and the dominance of its microbial food web on carbon cycling (Cotner et al. 2004). Lake Superior is, however, dimictic, thus there is density-driven complete vertical mixing of the water column in spring and early winter each year which homogenizes the water column. At present, available data indicate an imbalanced organic carbon budget in the lake. Estimates of community respiration of organic carbon of 13 to

84 Tg C per year far exceed the organic carbon inputs of 2.4 to 9 Tg C per year from atmospheric deposition, riverine flux and autochthonous primary production (Urban et al. 2005; Cotner et al. 2004). This mismatch in organic carbon input and output terms represents a major gap in the present understanding of the dynamics of the lake ecosystem (Sterner 2010), and is likely due to poor constraints on the respiration and the input estimates.

Amity Creek (in Duluth, MN), our most terrestrially-influenced site, drains a small, primarily forested basin. The creek flows into the Lester River less than 0.5 km before that river enters western Lake Superior. The area of Amity Creek watershed is 42.6 km$^2$. The watershed of the Creek is covered mostly by forest (71%) and grassland (19%) with little wetland (3%) and urbanized areas (2%) (www.lakesuperiorstreams.org).

### 4.2.2 Sampling

Samples were collected from Lake Superior during spring mixing in June, and thermal stratification in August of the year 2010 using the R/V *Blue Heron*. Site locations (Figure 4-1) and the sampling depths and coordinates are given in Table 1. Both surface (5 m) and deep water (127 m at WM, and 210 m at EM) were sampled at the offshore sites. The nearshore site (ONT) was sampled at only 4 m. Water samples were collected at each site using twelve 8 L Niskin bottles mounted on a rosette with Seabird model 911 plus Conductivity, Temperature, and Depth (CTD). DIC samples were taken directly from the Niskin bottles via pre-cleaned silicone tubing (10% v/v HCl, ACS Plus grade and then ultra-pure water [Millipore Milli-Q Plus]) into previously acid-cleaned and combusted (450°C for ≥ 4 hours) 0.5 L amber Pyrex bottles. After creating a known and consistent headspace, the samples were immediately preserved with saturated mercuric chloride solution, sealed air-tight with glass stoppers coated with Apiezon grease, and stored at room temperature in the dark until analysis. Bulk DOC and POC samples were obtained by filtering lake water through pre-combusted Whatman GF/F glass fiber filters (450°C for 4 hours; 0.7 μm nominal pore size) via nitrogen-pressurized stainless-steel canisters. Approximately 40 mL and 1.0 L of the bulk DOC was collected into an acid-cleaned and combusted amber glass vials and glass bottles, respectively, and each acidified to pH 2 using 6M HCl (ACS Plus grade) for concentration and radiocarbon analysis.

In September 2010, corn leaves (*Zea mays*) were collected from the watershed of western Lake Superior in order to determine the radiocarbon content of atmospheric $CO_2$. The sampling site was chosen to minimize fossil contamination (i.e., avoiding highways). The collected leaves were stored in perforated paper envelopes and refrigerated until analysis.

Amity Creek was sampled just above its confluence with the Lester River in September 2010 during baseflow condition. Creek water samples were filtered through pre-combusted Whatman GF/F glass fiber filters (450°C for ≥4 hours; 0.7 μm nominal pore size) to obtain DOC vs. POC. For DIC samples, the creek water was collected directly into previously acid-cleaned (10% HCl by volume, ACS Plus grade) and combusted 0.5 L amber Pyrex bottles after three rinses. After creating a known headspace, the sample was preserved with saturated mercuric chloride solution, air-tight sealed and stored at room temperature in the dark until analysis.

### 4.2.3 Ultrafiltration

Large-volume samples of 122-131 L and 168-223 L were drawn at the nearshore and offshore sites respectively, using an oil-free diaphragm pump and polypropylene tubing and fittings. At Amity Creek 80 L of creek water was drawn by hand with four 20-L stainless-steel canisters that had been previously pre-cleaned with soap and rinsed several times with ultra-pure water. In each case, the samples were filtered to remove bacteria and small particles using a cleaned (forward-and-back flushing with distilled water for ~4 hours, and rinsing with 7 L sample) dual layer (0.8 μm prefilter and 0.2 μm filter) polyether sulfone filter with polypropylene housing (Whatman Polycap 75 TC). Init DOC was obtained by sterile filtration (at 0.2 μm) to remove most bacteria and other POC from the total OC. Filter blank DOC, and TOC vs. Init DOC (cartridge filtered < 0.2 μm DOC) were monitored to ensure that the cartridge filters were not contributing measurable DOC to the samples. Portions (1 L) of Init DOC were collected in acid-cleaned and combusted 1 L bottles for radiocarbon and stable carbon isotope analysis. A stainless-steel and teflon cross-flow ultrafiltration system was used with a gear pump (Cole-Parmer Instrument Company, IL, USA) and an Amicon spiral-wound cellulose cartridge membrane (Amicon S10N1) with a nominal molecular weight cut-off of 1000 Da (based on >99% rejection of Vitamin $B_{12}$; Repeta et al. 2002, Repeta and Aluwihare 2006) to isolate HMW DOM (> 1000 Da) vs. LMW DOM (< 1000 Da). The ultrafiltration membrane was pre-cleaned by sequential rinsing with isopropanol, detergent (0.01% Micro-90), HCl (0.01 M), NaOH (0.01 M), and Milli-Q water, and then pre-conditioned with ~ 20 L of < 0.2 μm filtered sample water before processing the samples. LMW DOM was collected through the beginning, midway, and ending and the average concentration used in DOM mass balance analysis. The sample retentates were concentrated to 2-3 L; the resulting HMW DOM fractions were then frozen, and later freeze-dried in the lab for chemical and isotopic measurements.

### 4.2.4 DOC, DIC, POC, and PON measurements

Concentrations of DOC and DIC were measured on a Shimadzu $V_{CSH}$ analyzer. The analyzer was calibrated with potassium hydrogen phthalate (KHP) for DOC measurements, and primary standard grade sodium carbonate and ACS reagent grade sodium bicarbonate for DIC measurements (Zigah et al. 2011). As an additional reference check, deep seawater from Florida Strait at 700 m (Batch-8 2008) was obtained from the DOC Consensus Reference Program (Hansell laboratory, University of Miami, USA) and our measurements were 43.9 ± 3.9 μM (n = 5), in agreement with the consensus value of 41-43 μM. For each sample and standard, three injections were performed.

Particulate organic matter (POM) samples were freeze-dried and homogenized, fumigated with 12 N HCl (ACS Plus grade) in silver cups overnight to remove carbonates, dried again at 60°C, and cooled in a dessicator. They were then loaded into tin capsules and analyzed for particulate organic carbon (POC) and particulate organic nitrogen (PON) concentrations on a Costech ECS 4010 elemental analyzer (EA) coupled to Finnigan Delta Plus XP isotope ratio mass spectrometer (IRMS). The EA was calibrated with acetanilide (Costech Analytical Technologies Inc., Valencia, CA, USA) and caffeine (SigmalUltra; Wheaton Science Products, NJ, USA) reference standard materials. Typical instrumental precisions for POC and PON measurements were, 0.2% and 0.4% of the measured concentrations, respectively.

### 4.2.5 Radiocarbon ($\Delta^{14}C$) and stable isotope ($\delta^{13}C$ and $\delta^{15}N$) measurements

Radiocarbon and stable carbon isotope measurements were performed at the National Ocean Sciences Accelerator Mass Spectrometry Facility (NOSAMS) at the Woods Hole Oceanographic Institution, with the exception of the corn leaves sample, which was measured at the Keck Carbon Cycle AMS Facility (KCCAMS) at University of California, Irvine. DIC samples were directly hydrolyzed with $H_3PO_4$, and the resulting $CO_2$ was reduced to graphite with $H_2$ over Fe catalyst. Wet DOC, including bulk DOC and Init DOC, was converted to $CO_2$ by ultraviolet (UV) oxidation using a protocol based upon Beaupre et al. (2007) but oxidizing for three hours rather than four (Zigah et al. 2011). The evolved $CO_2$ was reduced to graphite. POC and freeze-dried HMW DOC samples were fumigated with 12 M HCl (ACS Plus grade) for 24 hours to remove carbonates, re-dried at 60°C, and combusted to $CO_2$ in a modified Carlo Erba NA1500 elemental analyzer, and then reduced to graphite. In all cases, a subsample of the $CO_2$ was taken for $\delta^{13}C$ measurement, and the graphite produced was compacted onto an aluminum cartridge target and analyzed by accelerator mass spectrometry (AMS) along with primary and secondary standards, and combustion and graphitization process blanks. The radiocarbon content of LMW DOC was calculated using a dual (HMW DOC and LMW DOC) isotopic mixing model as follows:

$$\Delta^{14}C_{\text{Init DOC}} = f \text{ x } \Delta^{14}C_{\text{ HMW DOC}} + (1\text{-}f) \text{ x } \Delta^{14}C_{\text{ LMW DOC}} \qquad (4\text{-}1)$$

where $f$ and (1- $f$) are the proportions of HMW DOC and LMW DOM, respectively, in the Init DOC. All radiocarbon values are reported as $\Delta^{14}C$, the part per thousand deviation of the sample's $^{14}C$:$^{12}C$ ratio relative to a nineteenth century wood standard that has been corrected to the activity it would have had in 1950 and a $\delta^{13}C$ of -25‰. $\Delta^{14}C$ was corrected for fractionation using $\delta^{13}C$ of samples according to the convention of Stuiver and Polach (1977). Precision of $\Delta^{14}C$ analysis is based on error of standards or multiple analyses on a target and ranged from 2-5‰.

Stable carbon isotope ratios ($\delta^{13}C$) of samples were measured at NOSAMS using an Optima stable isotope ratio mass spectrometer. Stable nitrogen isotope ratios ($\delta^{15}N$) were measured at the Large Lakes Observatory (LLO) using a Finnigan Delta Plus XP IRMS with Conflo III interface (Thermo Fisher Scientific Inc., Waltham, MA) coupled to a Costech ECS 4010 EA. The $\delta^{13}C$ of LMW DOC was also calculated from binary isotopic model as follows:

$$\delta^{13}C_{\text{ Init DOC}} = f \text{ x } \delta^{13}C_{\text{ HMW DOC}} + (1\text{-}f) \text{ x } \delta^{13}C_{\text{ LMW DOC}} \qquad (4\text{-}2)$$

Typical instrumental precision of $\delta^{15}N$ and $\delta^{13}C$ based on analyses of multiple external standards were 0.17‰ and 0.15‰, respectively. The stable isotope ratios ($^{13}C$:$^{12}C$ and $^{15}N$:$^{14}N$) are reported as $\delta^{13}C$ and $\delta^{15}N$ respectively, which are the per mil difference relative to Pee Dee Belemnite carbonate and atmospheric air standards. The $\delta$ notation is calculated as: $\delta Q = [R_{sample} / R_{standard} - 1] \times 1000$, where Q is either $^{13}C$ or $^{15}N$, and R is either $^{13}C$:$^{12}C$ or $^{15}N$:$^{14}N$.

### 4.2.6 $^{1}H$ Nuclear Magnetic Resonance (NMR) spectroscopy

All $^{1}H$ NMR spectra of HMW DOM were acquired using a Bruker Avance III 400 spectrometer. About 5 mg of each freeze-dried HMW DOM sample was dissolved in 1 ml of $D_2O$ (>99.9%, Aldrich Chemical Company, Milwaukee, WI) and analyzed in a 5 mm glass NMR tube (Wilmad Glass Co., NJ.). The solution-state $^{1}H$ NMR spectrum was acquired using a modified water suppression technique (Abdulla and Hatcher, in preparation) described by Lam and Simpson (2008), and a recycle delay of 2 s and 119 ms acquisition time.

### 4.2.7 Solid-state $^{13}C$ NMR

Solid-state NMR spectra were acquired by cross polarization/magic angle spinning (CP/MAS) using a Bruker Avance II 400 spectrometer operating at resonance frequency of 100 MHz for $^{13}C$. Rotor spin speed was 14 kHz and 4569 acquisitions were averaged for each spectrum. The optimum relaxation delay was 1 s, and the contact time was 1.5 ms. A 4 mm triple resonance probe was used. Chemical shifts are reported in 'δ' notation (in part per million, ppm) relative to tetramethylsilane.

## 4.3. Results

### 4.3.1 Concentrations of inorganic carbon and organic carbon size fractions

DIC was the largest C reservoir in the lake with a concentration of 820-829 μM during spring mixing in June (Table 4-2). In the thermally stratified lake in August, DIC concentration was 775-780 μM in the surface waters and increased to 805-806 μM in the deep waters (Table 4-3). DIC was also the largest C pool in Amity Creek with a concentration of 1869 μM (Table 4-3). Concentrations of OC in the lake were highest in the dissolved phases, with bulk DOC and Init DOC values of 93.8 ± 8.8 μM and 87.1 ± 10.1 μM, respectively, during spring mixing in June (Table 4-2). In the productive stratified lake in August, the concentration of bulk DOC and Init DOC was 97.1 ± 13.1 μM and 94.9 ± 10.4 μM, respectively (Table 4-3). Within the dissolved OC pool, LMW DOM was the largest fraction (76-92%) with concentration of 63.4-85.4 μM in the mixed lake in

June, and 74.2-85.1 μM during stratification in August (Table 4-2 and 4-3). HMW DOC concentrations of 5.8-11.0 μM in June and 11.3-27.3 μM in August correspond to 8-24% of DOC in the lake (Table 4-2 and 4-3). POC was the least abundant OC fraction with concentrations of 6.0-10.8 in the lake (Table 4-2 and 4-3). A similar trend in OC abundances was observed in Amity Creek with bulk DOC and Init DOC concentrations being the largest OC pools, followed by the LMW DOC (Table 4-2 and 4-3). The HMW DOC and POC concentrations were again considerably lower and similar to each other (POC of 24.5 μM vs. HMW DOC of 21.5 μM; Table 4-3). In general, the mean concentration within each OC size fraction was relatively higher in the surface waters during stratification compared to the values in the surface mixed lake (Table 4-2 and 4-3). The Init DOC was obtained by sterile filtration (at 0.2 μm) to remove most bacteria and other POC from the total OC. If the difference in bulk DOC (<GF/F) and Init DOC (<0.2 μm) is taken as the amount of bacterial carbon, then the bacterial carbon represents 2-7% of OC (and a substantial portion of POC) in the lake.

### 4.3.2 Elemental compositions of colloidal and particulate OC in Amity Creek and Lake Superior

The C and N content of HMW DOM in the lake (including both the offshore and nearshore samples) were 17.1 ± 1.8% (range 14.4-19.6%, n = 5) and 1.4 ± 0.1% (range 1.31-1.62%, n = 5), respectively in the mixed lake in June, and 20.9 ± 5.5% (range 15.2-29.9%, n = 5) and 1.50 ± 0.51% (range 1.10-2.37%, n = 5), respectively in the stratified lake in August (Table 4-1). The molar C:N value of HMW DOM (12.4-19.1) was consistently higher than concurrent POM (7.8-10.0) in the lake (Table 4-1). As with the %C, the C:N of HMW DOM was higher in the stratified lake relative to the mixed lake (except at EM 5 m) (Table 4-1). The HMW DOM in the Creek appears to be of poorer nutritional quality as it has a higher C content (29%) but lower N content (1.18%) than the lake samples. The C:N values of HMW DOM (C:N of 28.7) and POM (C:N of 11.2) in the creek was higher than the corresponding lake HMW DOM and POM, respectively (Table 4-1).

### 4.3.3 Stable isotopic distributions in DIC and size-fractionated OC in Lake Superior and Amity Creek

The $^{13}C$-enrichment of surface DIC relative to deep DIC was likely due to biochemical processes such as photosynthesis in the surface waters and respiration in the deep waters. In general, POC in the lake was more $^{13}C$-depleted (-29.0 ± 1.2 ‰) than concurrent bulk DOC (-26.4 ± 0.7‰), Init DOC (26.6 ± 0.8), HMW DOC (-26.9 ± 0.3‰) and LMW DOM (-26.5 ± 0.9‰) (Table 4-2 and 4-3). Similarly, POC in Amity Creek was $^{13}C$-depleted ($\delta^{13}C$-POC of -29.6‰) relative to the bulk, Init, HMW, and LMW DOC (Table 4-3). In the productive stratified lake in August, HMW DOC was relatively more $^{13}C$-depleted than the other dissolved OC size fractions (Table 4-3). $\delta^{15}N$ of POM ranged from -1.3‰ at ONT site to 2.1‰ in the deep waters at EM during spring mixing in June, and -4.1‰ to 2.8‰ in the stratified lake in August. The $\delta^{15}N$ of HMWDOM was 2.5 ± 0.6 ‰ (range 1.5-3.0‰, n = 5) in the isothermal lake in June, and 1.6 ± 1.3‰ (range 0.3-3.1‰, n = 5) in the stratified lake in August. HMW DOM was in general $^{15}N$-enriched relative to concurrent POM in Lake Superior (Table 4-1).

## 4.3.4 Radiocarbon distributions within OC size fractions in Lake Superior and Amity Creek

DIC in the lake had a modern (bomb) radiocarbon signature with $\Delta^{14}C$ of 52 ± 10‰ (range 39-69‰, n = 10) (Table 4-2 and 4-3). The $\Delta^{14}C$ of Init DOC of 44 ± 15‰ (range 17-59‰, n = 10) was similar to the $\Delta^{14}C$ of HMW DOC of 49 ± 14‰ (range 23-64‰, n = 10) and the $\Delta^{14}C$ of LMW DOC of 44 ± 16‰ (range 17-62‰, n = 10) (Table 4-2 and 4-3). As with DIC, Init DOC and its constituent HMW and LMW DOC consistently reflected a modern $^{14}C$ signature. Throughout sampling (with the exception of WM site during isothermal lake), the $\Delta^{14}C$ of DIC was comparable to those of Init, HMW, and LMW DOC in the lake (Table 4-2 and 4-3). The radiocarbon content of POC was more variable in the lake. In the stratified lake in August, surface POC ($\Delta^{14}C$ of 37-53‰) was modern, and slightly $^{14}C$-depleted relative to concurrent DIC ($\Delta^{14}C$ of 57-65‰). In contrast, deep POC ($\Delta^{14}C$ of 12‰ at WM and -94‰ at EM) in the stratified lake was more $^{14}C$-depleted than concurrent DIC ($\Delta^{14}C$ of 56‰ at WM and 69‰ at EM) (Table 4-3).

During spring mixing in June, POC had a $\Delta^{14}C$ of 14 ± 27‰ (range -15 to 48‰), and except for POM from deep water at the EM site, was consistently more $^{14}C$-depleted than co-occurring DIC (44 ± 4‰) (Table 4-2). Also, POC was consistently more $^{14}C$-depleted than concurrent Init, HMW, and LMW DOC during spring mixing in June (Table 4-2). Contrary to the other dissolved OC fractions, the radiocarbon signature of bulk DOC was more variable with a $\Delta^{14}C$ of 30 ± 28‰ (range -19 to 57‰, n = 10) in the lake. Bulk DOC at the nearshore ONT site contained a pre-bomb radiocarbon signature (pre-aged) with $\Delta^{14}C$ values of -19 ± 5‰ (90 ± 40 years BP) and -8 ± 5‰ (40 ± 5 years BP), respectively during mixed and stratified lake conditions (Table 4-2 and 4-3). Offshore bulk DOC was comparable to concurrent DIC and the other dissolved OC fractions, except at the EM site during stratification, where the surface and deep bulk DOC (as with the POC at this site) was $^{14}C$-depleted relative to concurrent DIC and other dissolved OC fractions (Table 4-2 and 4-3).

DIC in Amity Creek during baseflow conditions in September 2010 had $\Delta^{14}C$ of 42 ± 3‰, similar to the $\Delta^{14}C$ of atmospheric $CO_2$ (37 ± 2‰) as determined from a corn plant taken in the same year (Table 4-3). $\Delta^{14}C$ values of HMW DOC (36 ± 4‰) in the creek were similar to concurrent DIC (and atmospheric $CO_2$), but more $^{14}C$-enriched than Init and LMW DOC, and POC (Table 4-3). The oldest OC size fraction in the creek (and lake samples) was POC from Amity Creek ($\Delta^{14}C$ of -127 ± 4‰, 1040 ± 35 BP years) (Fig. 4-3). A cross plot of pooled $\Delta^{14}C$ of DIC vs. $\Delta^{14}C$ of OC size fractions shows that most of the OC fractions in the lake are largely constrained to $\Delta^{14}C$ of 40-60‰; $\Delta^{14}C$ of bulk DOC and POC showed greater spread in $\Delta^{14}C$ values than the other OC fractions (Fig. 4-4). We could not perform statistical analysis on the $\Delta^{14}C$ values because they remained skewed even after log transformation.

## 4.3.5 $^{13}C$ NMR data

The $^{13}C$ NMR spectra of the HMW DOM from the various lake sites displayed similar chemical shifts and resonances (Fig. 4-5). The $^{13}C$ NMR spectra were characterized by four major resonances at 180 ppm, 110 ppm, 70 ppm, and 25-30 ppm (Table 4-4), corresponding to peaks from carbon in the functional groups COOH, OCO, HCOH, and $CH_2/CH_3$, respectively (Repeta et al. 2002, Abdulla et al. 2010). Quantification of the NMR spectra shows O-alkyl carbohydrate carbon (HCOH) was the dominant component of HMW DOC, and constitutes 49 ± 5% of the HMW DOM in Lake Superior. Paraffinic carbon, $CH_x$ (14 ± 1%) and carboxyl and aliphatic amide carbons, COO/CON (14 ± 3%) were also significant components of HMW DOC (Table 4-4). Substituted aromatic carbon or double bonded carbon (C=C/Ar-C), oxygen substituted aromatic carbon (Ar-O), and carbonyl carbon were each only trace components (≤ 2%) of the HMW DOC in the water column of Lake Superior (Table 4-4).

For compound class information, the spectra were integrated over the regions 0-60 ppm for aliphatic carbon, 60-120 ppm for O-alkyl and anomeric carbohydrate carbon, and 120-160 ppm for aromatic carbon (Bianchi et al. 2004, Koprivnjak et al. 2009, Abdulla et al. 2010). The NMR data show that the total carbohydrate fraction was 60 ± 5%, and was the largest biochemical compound class within the HMW DOM in the lake (Table 4-4). The highest amount of total carbohydrates (70%) was observed in the stratified surface water at site EM. During stratification, the amount of total carbohydrates decreased slightly from 61% in the surface water to 55% in the deep water at the WM site (Table 4-4). Total aliphatic carbon made up 21 ± 2% of HMW DOC, whereas aromatic carbon was the least abundant compound class (4 ± 1%) within the HMW DOC in Lake Superior (Table 4-4). A cross plot of compound class composition vs. $\Delta^{14}C$ of HMW DOM appears to suggest that aliphatic content increases with $\Delta^{14}C$; that is addition of aliphatic components yields $^{14}C$-enriched radiocarbon values (Fig. 4-6). In contrast, $\Delta^{14}C$ of HMW DOM were not modulated by changes in the carbohydrate and aromatic contents (Fig. 4-6).

## 4.3.6 $^{1}H$ NMR data

The $^{1}H$ NMR spectra of all the samples (Fig. 4-7) show characteristic major resonances at 3.3 ppm from carbohydrate protons (CHOH), 2.0 ppm from methyl protons of acetate ($CH_3CO$), and an intense peak at 1.1 ppm from methyl and methylene protons ($CH_3$ and $CH_2$) of lipids (Aluwihare et al. 1997, Aluwihare et al. 2002, Repeta et al. 2002). In contrast to open-water Lake Superior, HMW DOC from Amity Creek, the most terrestrially-influenced site, shows a broad peak at 6-9 ppm from aromatic protons (Fig. 4-7A) (Malcolm 1990, Abdulla et al., Unpubl.) and the nearshore site ONT shows a similar, but smaller broad aromatic peak.

The $^{1}H$ NMR spectra were integrated (Table 4-5) to obtain the relative compositions of biochemical compounds such as total carbohydrates (0.95-1.33 ppm from protons of the methyl group of the 6-deoxy sugars, and 3.19-5.80 ppm from protons of O-alkyl carbohydrate, HCOH), bound acetates (1.76-2.03 ppm from protons of $CH_3CO$), and aromatic compounds (5.80-9.00 ppm from aromatic protons) in the HMW DOM (Repeta et al. 2002, Koprivnjak et al. 2009). These integrations show that acetates and aromatic compounds account for 11 ± 1%, and 4 ± 1% of the HMW DOM. The carbohydrate composition of HMW DOM (37 ± 4%) as indicated by the $^{1}H$ NMR data (Table 4-5) is lower than the value from $^{13}C$

NMR (Table 4-4). This is because the water suppression technique employed to optimize the NMR signal by suppressing the more abundant protons also attenuates the carbohydrate signals in the samples by ~ 30% (Lam and Simpson, 2008). Although the actual values differ, the basic trends in biochemical compositions as indicated by the both $^{1}H$ and $^{13}C$ NMR remained similar (Table 4-4 and 4-5). Amity Creek HMW DOM, which is relatively more terrestrially influenced, shows higher resonances from aromatic carbon (9%) than the Lake Superior nearshore ONT site (5-7%), and offshore sites (3-4%) (Table 4-5). The HMW DOM from Amity Creek also had lower total carbohydrate protons (26%) and acetate (9%) content relative to the Lake Superior sites (Table 4-5).

## 4.4 Discussion

### 4.4.1 Abundances and distributions of size-fractionated OC in Lake Superior

The abundance of DIC in the surface waters decreased during stratification likely due to both increased conversion of inorganic carbon into organic molecules through photosynthesis, and lower $CO_2$ solubility in warmer surface water (Zigah et al. 2011). It was interesting to observe that all the OC phases including the particulate (POM), dissolved (bulk DOM and Init DOM), colloidal (HMW DOM), and LMW DOM did increase in their relative abundances in the surface waters (Table 4-3) following addition of freshly synthesized materials from photosynthesis. This suggests that some portion of the lake's primary production is routed through the dissolved pool, consistent with observations in the world's oceans (Azam et al. 1983; Hedges 1992; Urban et al. 2005; Kaiser and Benner 2009; Mannino and Harvey 2000). The relatively higher OC abundances at the nearshore (ONT) site could be due to higher primary production supported by terrestrially-derived nutrients and increased nearshore temperatures (Urban et al. 2005; Zigah et al. 2011). The WM site also exhibited relatively higher abundances in dissolved, colloidal, and low molecular weight OC during stratification than the EM site (Table 4-2 and 4-3), possibly because of the relatively higher primary productivity in this region of the lake (McManus et al. 2003; Cotner et al. 2004).

The proportion of HMW DOC in the lake was 8-24% (Table 4-1) implying LMW organic molecules dominate the OC in the lake. Our result is consistent with Repeta et al. (2002) and Urban et al. (2005), who in a study at the far western arm, and just off the Keweenaw Peninsula in Lake

Superior noted that the LMW DOC was the largest pool of the bulk OC, although these studies reported larger percentages of HMW DOC (34-50%), most likely due to the relative proximity of their respective sites to shore. The molecular weight distribution in Lake Superior is similar to that seen in open oceans; most oceanic DOM (60-80%) is LMW (Guo et al. 1995; Amon and Benner 1996; McCarthy et al. 1996; Hansell and Carlson 2002; Repeta et al. 2002; Walker et al. 2011). In contrast to Lake Superior and the oceans, estuarine DOM (13-48% of HMW DOM; Sigleo 1996; Guo et al. 1996; Wang et al. 2004) and riverine DOM (24-87% HMW DOM; Benner and Hedges 1993; Hedges et al. 1994; Benner and Opsahl 2001; Repeta et al. 2002) contain a higher proportion of HMW molecules because of relatively greater terrestrial influence and higher aquatic primary production.

### 4.4.2 Partitioning of C and N in HMW DOM and Bulk POM: Insights into source and diagenetic state of colloidal and particulate OM

The elemental N and C compositions of colloidal and particulate OM can be useful in deciphering/inferring the source and degree of diagenesis of these OC size classes in aquatic systems. Organic materials that have undergone diagenetic alterations usually have high C:N values because preferential release of N relative to C yields higher C:N values in the residual mineralized material after microbial degradation (Hopkinson et al. 1997; Abell et al. 2000; McCallister et al. 2006). Also, OM source classification based on C:N is possible because terrestrial vascular plants (and by extension, terrestrial DOM) typically have much higher C:N values ($\geq$ 21; Hedges et al. 1997) than fresh aquatic algal-derived OM (~6.6; Redfield et al. 1963).

In Lake Superior, the molar C:N values of HMW DOM (12.4-19.1) were consistently higher than those of concurrent POM (7.8-10) (Table 4-1) in agreement with previous studies by Urban et al. (2004, 2005) who reported colloidal C:N of 25.7 ± 2.6 vs. POM C:N of 8.5 ± 0.3 off the Keweenaw Peninsula of Lake Superior. The relatively higher colloidal C:N values reported by Urban et al. (2005) could be due to higher primary production and terrestrial influence because of greater relative proximity to shore than for our offshore study sites. Also, McCallister et al. (2006) in their study of the York River estuary reported HMW DOM C:N values of 7.4-17.9 consistently higher than POM C:N of 5.5-9.0. The higher C:N of HMW DOM relative to C:N of POM in Lake Superior could be due to higher microbial reworking, and/or other diagenetic alteration (i.e., photochemical or physical) of the HMW DOM (Abell et al. 2000; McCallister et al. 2006; Guo and Macdonald 2006), or different sources of HMW DOM vs. POM (Hedges et al. 1986). The higher C:N values of HMW DOM compared to POM could be reflecting different sources with relatively higher terrestrial influence of HMW DOM (Hedges et al. 1986; Lobbes et al. 2000; Wang et al. 2004; Urban et al. 2005; McCallister et al. 2006). Also, greater concentration of dissolved free amino acids and amino sugars in the LMW DOM is possible because these compounds could have molecular weight lower than the 1 kDa cut off for HMW DOM thereby depleting the N content of HMW DOM and elevating its C:N ratio (Amon and Benner 1996; Hansell and Carlson 2002; Guo and Macdonald 2006). It is, however, worth noting that these dissolved organic nitrogen compounds are typically only a trace constituent of DOC, and their loss is likely not the cause of the elevated HMW DOM C:N values seen in Lake Superior (Hansell and Carlson 2002). Further, selective sorption of N to especially fine POM could potentially decrease POM C:N values relative to C:N of HMW DOM (Hedges et al. 1986; Aufdenkampe et al. 2001; McCallister et al. 2006).

The C:N values of HMW DOM from Lake Superior (12-19) are consistent with C:N values of HMW DOM from various oceanic sites including the Pacific (15-23), and Atlantic (16-18) Oceans (Amon and Benner 1994; McCarthy et al. 1996; Benner et al. 1997) and estuarine HMW DOM of 9.6-24.5 (Guo et al. 1996; Engelhaupt and Bianchi 2001; Wang et al. 2004; McCallister et al. 2006), but lower than riverine HMW DOM C:N of 34-87 (Amon and Benner 1996; Repeta et al. 2002). Higher riverine HMW DOM C:N values most likely reflect higher terrestrial influence, consistent with the relatively higher HMW DOM C:N values at Amity Creek (~29: Table 4-1) (our most terrestrially impacted site). In Lake Superior, the shift to higher C:N values of HMW DOM and POM during stratification (Table 4-1) is possibly reflecting higher accumulation of fresh lipid compounds in these OC sizes from increased primary production, and/or enhanced microbial and/or photochemical mineralization in the warmer stratified lake in August.

### 4.4.3 Sources and turnover times of OC size classes based on radiocarbon and stable isotopic distributions

Natural abundance stable isotopes ($\delta^{13}C$ and $\delta^{15}N$) could provide source distinction and diagenetic state information on OC, and the concurrent use of $\Delta^{14}C$ which is a source, process, and time tracer would not only help to accurately constrain OC sources, but also infer their residence times in aquatic environments. We analyzed $\Delta^{14}C$ and $\delta^{13}C$ (and also $\delta^{15}N$ in the case of HMW DOM and POM) of the OC size fractions including the particulate, dissolved, colloidal, and low molecular weight fractions in Lake Superior in order to examine their potential sources, residence times, and diagenetic state and/or fate.

While primary production (9.73 Tg C $yr^{-1}$; Sterner 2010) is by far the dominant source of OC to the lake, most of this does not persist in the lake, but rather is lost through biochemical degradation or respiration by heterotrophic microbes (Cotner et al. 2004; Urban et al. 2005). Mair and Swain (1978) estimated that almost all newly added DOC from phytoplankton production would not persist over a year in the lake whereas Urban et al. (2005) estimated that autochthonous DOC recycles within 8 years in the lake. These rapid turnover times of DOC are consistent with the modern $\Delta^{14}C$ values of the DOC fractions seen in the lake (Table 4-2 and 4-3).

The general $^{15}N$-enrichment of HMW DOM could be due to microbial and/or diagenetic alterations because $^{14}N$ is preferentially released during remineralization (Lehmann et al. 2002). The general $^{15}N$-enrichment of HMW DOM (relative to POM), together with the elevated C:N values, and modern radiocarbon signatures suggest that the lake HMW DOM is both rapidly altered and rapidly recycled. That HMW DOM is of contemporary photosynthetic origin and recycles rapidly has also been observed in several oceanic systems based on radiocarbon (Santschi et al. 1995; Guo et al. 1996; Loh et al. 2004; Aluwihare et al. 2002; Repeta and Aluwihare 2006), molecular and/or biochemical composition (McCarthy et al. 1996; Skoog and Benner 1997; Mannino and Harvey 2000; Aluwihare et al. 2002; Loh et al. 2004) and bacterial utilization of size-fractionated DOC (Amon and Benner 1994, 1996). The $^{14}C$-depletion of HMW DOC relative to lake DIC in the stratified surface waters at WM is intriguing, and could suggest a unique source such as that year's terrestrially-derived colloids or sedimentary colloids (Guo et al. 1996; Guo and Macdonald 2006).

Most of the lake's DOC is present as LMW DOC. Since the fraction of freshly added autochthonous DOC is relatively small, and recycled rapidly, it is generally thought that most of the DOC in the lake, which is present as LMW DOC, is derived from terrestrial DOC (Mair and Swain 1978; Urban et al. 2005). The modern radiocarbon signature of LMW DOM in the lake indicates a semi-labile nature, and also suggests that most of the terrestrial DOC coming into the lake is from contemporary terrestrial litter rather than from old deep horizon soil OC. The semi-lability of LMW DOM in Lake Superior puts the turnover time of LMW DOC at ≤ 60 years, much shorter than the residence time (170 years) of the water in the lake. This is different from observations in oceanic systems where the LMW DOC is typically pre-aged and recycles on a longer time scale than the Init DOC and HMW DOC (Amon and Benner 1994; Guo et al. 1995; Santschi et al. 1995; Skoog and Benner 1997; Mannino and Harvey 2000; Loh et al. 2004). The semi-lability of LMW DOC in Lake Superior relative to the oceans is likely due to the dimictic nature of the lake, which overturns completely twice per year, bringing deep water DOC to the more reactive (both microbially and photochemically) surface waters at a regular and faster rate than can occur in the open ocean. The lake also has a shorter water residence time (170 years) than seen for the deep ocean (1000-1500 years).

$^{14}$C-depletion of bulk DOC relative to DIC and other OC size classes in the stratified waters at the EM (both surface and deep waters) and ONT sites suggest a different source, some sedimentary influence and/or preferential remineralization of younger DOC. The consistent $^{13}$C-depletion of POC relative to the dissolved phases of OC, and the $^{14}$C-depletion of POM relative to concurrent DIC (in some cases) is consistent with a unique and old source, most likely, resuspended sedimentary materials. Several previous studies in the lake have reported resuspension events in different parts of the lake and noted some variable portions of the lake POC are derived from resuspended materials (Baker and Eisenreich 1989; Hawley 2000; Urban et al. 2004; Zigah et al. 2011).

Our baseflow data from Amity Creek show that HMW DOM ($\Delta^{14}$C of 36 ± 4‰) was similar to the creek DIC ($\Delta^{14}$C of 42 ± 3‰), suggesting the creek HMW DOM is probably bio-reactive with a rapid turnover time. While the baseflow bulk and Init DOC $\Delta^{14}$C show $^{14}$C-depleted DOC is being drained from the watershed by the creek, highflow DOC from the creek is more $^{14}$C-enriched implying that the large amount of DOC delivered from the creek during flow events is modern (Zigah et al. 2011). LMW DOM from baseflow condition was pre-aged ($\Delta^{14}$C of -5 ± 4‰; Table 4-3) indicating older sources or longer residence time in the creek.

### 4.4.4 Total carbohydrate abundance and bio-reactivity of HMW DOM in Lake Superior

The CPMAS $^{13}C$ NMR and $^{1}H$ NMR data indicate that carbohydrate is a major biochemical compound-class within HMW DOM (33-70%) in Lake Superior (Table 4-4), consistent with previous observations in the lake (Repeta et al. 2002). Carbohydrate content in HMW DOM in Lake Superior agrees with reported carbohydrate content of oceanic HMW DOM (40-70%) as determined by $^{13}C$-NMR (Benner et al. 1992; McCarthy et al. 1993), $^{1}H$-NMR (Aluwihare et al. 2002; Repeta et al. 2002; Gogou and Repeta 2010), molecular-level aldoses (McCarthy et al. 1993; 1996, Skoog and Benner 1997; Aluwihare et al. 2002; Repeta and Aluwihare 2006), acid hydrolysis (Burney and Sieburth 1977; Loh et al. 2004), and 3-methyl-2-benzothiazoninone hydrazine (MBTH) assays/calorimetric detection method (Burney and Sieburth 1977; Benner et al. 1992; Pakulski and Benner 1992). Similarly, carbohydrates were the dominant biochemicals in HMW DOM from rivers based on proton and $^{13}C$ NMR (Repeta et al. 2002; Bianchi et al., 2004), and in estuaries based on $^{13}C$ NMR (Van Heemst 2000) and acid hydrolysis (Wang et al. 2004).

That carbohydrates are the largest component of colloidal OC in Lake Superior (and other aquatic systems) is not surprising because carbohydrates are the most abundant identified fraction of OM in freshwaters (Tranvik and Jorgensen 1995; Repeta et al. 2002), estuaries (Sigleo 1996; Wang et al. 2004), and in the oceans (Benner et al. 1992; McCarthy et al. 1996; Wang et al. 1998; Mopper et al. 2007). In a study in the Gulf of Mexico and North Pacific Ocean, McCarthy et al (1993) reported that carbohydrates constituted ~50% of total carbon in the HMW DOM. The abundance of carbohydrates in these aquatic environments could be linked to their roles as structural (e.g. chitin and peptidoglycan), and energy storage (e.g. glycogen) components of aquatic living organisms, being a major component of the cell and/or cell wall of algae (up to 70% of carbohydrate; Romankevich 1984), bacteria (Boon et al. 1998; McCarthy et al. 1998), fungi (Wershaw et al. 2005), and terrestrial plants (up to 65% of carbohydrates mostly as cellulose and hemicellulose) (Mannino and Harvey, 2000; Wershaw et al., 2005). In a study at the Northeastern Pacific Ocean, Wang et al. (1998) reported that carbohydrate-like material made up the largest fraction of phytoplankton in the surface waters. Similarly, Haug et al. (1973) in a study in Trondheimsfjord, an inlet of the Norwegian Sea found that dinoflagellates contained greater abundance of carbohydrates than the other biochemicals.

The lower relative abundance of carbohydrates in the deep water at WM site in Lake Superior corresponds to lower abundances of HMW DOM at depth (Table 4-1), and is consistent with the idea that carbohydrate is a significant portion of the contemporary and labile HMW DOM (Benner et al. 1992; Bianchi et al. 1995). Lower deep-water carbohydrate concentration in Lake Superior is similar to observations in several oceanic systems (Benner et al. 1992; Pakuski and Benner 1994; Guo et al. 1995; Bhosle et al. 1998; Borch and Kirchman 1997; Aluwihare et al. 2002), and probably reflects microbial mineralization of the carbohydrates in the water column due to its bio-reactivity. Carbohydrate composition of HMW DOC was not related to $\Delta^{14}C$ of HMW DOC (Fig. 4-6) probably because of multiple sources of HMW DOC in the lake such as terrestrial and algal-derived colloids.

### 4.4.5 Potential origin of HMW DOM in Lake Superior based on NMR spectroscopy

In Lake Superior, both $^{1}H$ NMR and $^{13}C$ NMR spectra show that the HMW DOM is predominantly aliphatic (19-46%), thus containing more aliphatic (e.g. lipid and protein) components with little aromatic contribution (2-7%) (e.g. lignophenols). While some studies have attributed the higher proportion of aliphatic structures in the HMW DOM to reflect algal origin (Benner et al. 1992; Hedges et al. 1992; Bianchi et al. 1995; Bianchi et al. 2004), others have reported a high degree of similarity of HMW DOM from more terrestrially influenced (creeks, rivers, some lakes) and less terrestrially influenced aquatic systems (McKnight et al. 1997; Repeta et al. 2002). In Lake Superior, although some amount of algal-derived DOC is added to the lake each year (Cotner et al. 2004; Urban et al. 2005; Sterner 2010), this small fresh DOC is more bio-reactive and is mineralized rapidly. In contrast, most of the semi-labile bulk DOC in the lake is derived from terrestrial sources (Urban et al. 2005), hence it is likely that the HMW DOM is derived from both algal and terrestrial sources, with the algal portion being most labile portion of HMW DOM. The presence of pre-aged lipids, and some pre-aged amino acids and non-hydrolyzable organic fractions in the lake HMW DOM (Chapter five) is consistent with multiple sources of components in the HMW DOM. The lower aromatic signature of the HMW DOM from the lake could be due to photo-bleaching or photochemical alterations which lead to losses of dissolved phenolic compounds from lignin and tannins, thereby leading to an underestimation of terrestrial colloidal inputs (Opsahl and Benner 1998; Opsahl and Zepp 2001; Minor et al. 2007).

The $^{1}H$ NMR spectra show that HMW DOM from open-water Lake Superior is relatively rich in acetate (10-14%; Table 4-5) compared to a nearshore Lake Superior site (3%; Repeta et al. 2002), Amity Creek (9%; Table 4-5), and a suite of lakes and ponds (3%; Repeta et al. 2002) which are more terrestrially influenced. Our data are consistent with observations in the oceans where the HMW DOM is relatively rich in acetate functional groups from amino sugars (10%; Repeta et al. 2002), and suggest some microbial input (either aquatic or soil bacterial cells) to the HWM DOM in the lake. Several studies in the oceans and estuaries show bacteria could provide substantial carbon sources to the HMW DOM pool (Koike et al. 1990; McCarthy et al. 1998; Ogawa et al. 2001; Benner and Kaiser 2003; Zou et al. 2004). Minor et al. (2001) reported refractory bacterial components in the HMW DOM from Delaware Bay, and Zou et al. (2004) also noted reported bacterial-derived old lipid materials in the HMW DOM from some estuaries.

## 4.5 Conclusion

In the more productive stratified lake, surface water abundances of DOC size fractions increased, indicating addition from algal-derived DOC to the surface. The stratified surface waters make up a small fraction ($\leq$ 5%; Mair and Swain 1978; Cotner et al. 2004; Urban et al. 2005) of the lake, and since the deep water abundances of DOC size fractions did not exhibit increases during stratification, we suspect that majority of the DOC in the lake is derived from terrestrial sources potentially through accumulated riverine inputs. HMW DOM was relatively N-poor (C:N of 12-19) compared to POM (C:N of 8-10) revealing either a more diagenetically altered state or a different source function than POM in the lake.

Coupled together, $\Delta^{14}C$, $\delta^{13}C$, $\delta^{15}N$, and C:N signatures together robustly constrain HMW DOM in Lake Superior as younger, mostly from recent fresh algal and/or terrestrial photosynthetic origin, bio-reactive, and most of it recycles rapidly ($\leq$ 3 years). In contrast, these multi-isotopic and elemental signatures show that POM in the lake was generally older (especially in the mixed lake), includes an older, most likely resuspended-sediment source rather than recent fresh algal and/or terrestrial production. $^{13}C$ NMR and $^{1}H$ NMR show that HMW DOM in Lake Superior has more acetate, and aliphatic compounds (19-46%) with only trace aromatic signature (2-7%), indicating algal and microbial sources and possibly photochemically and microbially altered terrestrial material.

$\Delta^{14}C$ values of LMW DOC were similar to $\Delta^{14}C$ of concurrent DIC. The $\Delta^{14}C$ and $\delta^{13}C$ composition of LMW DOM suggests that most of the LMW DOM in the lake is derived from diagenetically altered terrestrial DOC, and is semi-labile with turnover time of a few decades ($\leq$ 60 years). The semi-lability of LMW DOC in Lake Superior is attributed to the dimictic nature of the lake, where overturn of the lake water each spring and fall brings deep-water DOC to the more reactive surface waters, thus leading to faster, more efficient DOC remineralization than can occur in the open ocean.

While HMW DOM is younger and more bio-reactive than LMW DOM, we are not able to extend this size-age relationship to include the particulate phase (POM) as has been done in oceanic systems because POM in the lake appears to strongly affected by sediment resuspension. POC ages in most riverine systems have also been shown to be much older than concurrent DOC due to old sources (such as sedimentary rocks and fossil-derived carbon) and long term storage in stream and riverine sediment beds. The increases in the abundance of POM in the surface water relative to deep water during thermal stratification imply that some fresh POM is added from algal production. The $\Delta^{14}C$ and $\delta^{13}C$ of bulk/composite POM in the lake in some cases did not reflect the modern fresh POM source, implying the freshly added POM does not persist, but rather is preferentially recycled rapidly. It is recommended that future studies isolate the fresher POM, for instance through flow cytometric sorting, and determine its isotopic and chemical composition to allow for determination of its bio-reactivity and turnover time.

In Amity Creek, baseflow colloidal OM had a modern radiocarbon signature indicating recent terrestrial photosynthetic origin, relatively higher bio-reactivity, and rapid turnover time. Pre-aging of the creek LMW DOC relative to creek colloids, contrary to observation in Lake Superior, could be due to input from an old source, perhaps deeper soil horizons within the watershed. Also, creek colloidal OM was more N-depleted (elevated C:N value) than the lake colloids due to relatively higher terrestrial influence. POC in the creek was pre-aged (1040 years BP) and the POC-DIC $\Delta^{14}C$ offset indicates most of the POC originated from older particulates stored in the soil (or from deeper horizon) or streambed for longer time period.

**Table 4-1. Details of sampling stations and depths, ultrafiltration, and the elemental carbon and nitrogen compositions of HMW DOM and bulk POM. The values in parenthesis are the standard deviations from analyses of duplicate natural samples.**

| Site description | Latitude (°N) | Longitude (°W) | Depth sampled (m) | Water depth (m) | Ultrafiltration | | | | | HMW DOM | | | | POM | |
|---|---|---|---|---|---|---|---|---|---|---|---|---|---|---|---|
| | | | | | Total vol. (L) | Retentate vol. (L) | CF | HMW DOM recovery (%) | Mass balance (%) | %C | %N | Molar C:N | $\delta^{15}N$ (‰) | Molar C:N | $\delta^{15}N$ (‰) |
| June 2010 Isothermal | | | | | | | | | | | | | | | |
| ONT | 46°54 | 89°20 | 5 | 20 | 131 | 2.2 | 60 | 12 | 86 | 17.0(0.2) | 1.31( 0.01) | 15.1(0.3) | 1.5(0.3) | 9.1 | -1.3 |
| WM | 47°19 | 89°51 | 5 | 171 | 203 | 2.0 | 102 | 8.4 | 101 | 17.0(0.3) | 1.42(0.04) | 13.9(0.6) | 2.5(0.2) | 7.8 | 0.9 |
| WM | 47°19 | 89°51 | 127 | 171 | 214 | 2.0 | 107 | 8.5 | 88 | 17.3(0.5) | 1.62 (0.04) | 12.46(0.03) | 2.9(0.0) | 8.4 | 1.1 |
| EM | 47°34 | 86°39 | 5 | 242 | 213 | 1.5 | 142 | 12 | 92 | 19.6(0.6) | 1.34(0.04) | 17.0(0.1) | 2.4(0.1) | 8.5 | 1.7 |
| EM | 47°34 | 86°39 | 210 | 242 | 192 | 2.3 | 83 | 10.3 | 89 | 14.4(0.1) | 1.35(0.02) | 12.4(0.1) | 3.0(0.1) | 8.2 | 2.1 |
| August 2010 Stratified | | | | | | | | | | | | | | | |
| AC | 46°84 | 92°01 | < 1 | nd | 80 | 1.6 | 50 | 21.2 | 94 | 29.0(0.2) | 1.18 (0.02) | 28.7(0.1) | 1.5(0.0) | 11.2 | 2.6 |
| ONT | 46°54 | 89°20 | 5 | 20 | 122 | 2.8 | 44 | 24.3 | 108 | 18.0(0.3) | 1.10 (0.01) | 19.1(0.2) | 0.3(0.3) | 9.6 | -2.6 |
| WM | 47°19 | 89°51 | 5 | 171 | 175 | 2.1 | 83 | 19.4 | 104 | 20.5(0.1) | 1.40 (0.03) | 17.0(0.3) | 1.7(0.1) | nd | nd |

| | | | | | | | | | | | | | | |
|---|---|---|---|---|---|---|---|---|---|---|---|---|---|---|
| WM | 47°19 | 89°51 | 127 | 171 | 168 | 1.6 | 105 | 14.3 | 93 | 21.0(0.5) | 1.5(0.1) | 16.2(0.2) | 2.6(0.2) | 9.5 | 2.8 |
| EM | 47°34 | 86°39 | 5 | 242 | 223 | 1.2 | 186 | 12.3 | 82 | 29.9(0.1) | 2.37(0.01) | 14.74(0.01) | 0.4(0.1) | 10.0 | -4.1 |
| EM | 47°34 | 86°39 | 210 | 242 | 205 | 2.5 | 82 | 13.2 | 105 | 15.15(0.04) | 1.14 (0.02) | 15.5(0.3) | 3.1(0.2) | nd | nd |

CF is concentration factor and is calculated as total volume divided by retentate volume.
HMW DOM recovery = DOC concentration of retentate multiplied by the retentate volume, all divided by the concentration of Init DOC multiplied by the init volume.

**Table 4-2. Concentration, stable carbon, and radiocarbon isotopic compositions of size fractionated OM and DIC in isothermal Lake Superior in June 2010.**

| | *Concentration* *(µM)* | $\delta^{13}C$ *(‰)* | $\Delta^{14}C$ *(‰)* | $^{14}C$ *ages* *(BP yrs)* | *NOSAMS* *Accession #* |
|---|---|---|---|---|---|
| *ONT 4 m* | | | | | |
| DIC | 820 ± 2 | 0.7 | 48 ± 4 | Modern | OS-82065 |
| POC ( > 0.7 µm) | 10.9 | -29.8 | -10 ± 4 | 25 ± 30 | OS-82067 |
| Bulk DOC (< 0.7 µm) | 108.7 ± 1.9 | -28.2 | -19 ± 5 | 90 ± 40 | OS-82423 |
| Init DOC (< 0.2 µm) | 92.3 ± 0.5 | -26.0 | 52 ± 4 | Modern | OS-82503 |
| HMW DOC (>1kDa) | 10.6 ± 1.6 | -27.2 | 34 ± 4 | Modern | OS-88673 |
| LMW DOC (< 1kDa) | 81.7 | -25.8 | 54 ± 4 | Modern | |
| *WM 5 m* | | | | | |
| DIC | 830 ± 4 | 0.4 | 46 ± 4 | Modern | OS-82054 |
| POC ( > 0.7 µm) | 4.8 | -29.7 | -15 ± 2 | 65 ± 15 | OS-83649 |
| Bulk DOC (< 0.7 µm) | 87.9 ± 0.6 | -26.0 | 57 ± 4 | Modern | OS-82188 |
| Init DOC (< 0.2 µm) | 69.2 +0.8 | -28.3 | 17 ± 4 | Modern | OS-82424 |
| HMW DOC (>1kDa) | 5.8 | -27.3 | 23 ± 5 | Modern | OS-88747 |
| LMW DOC (< 1kDa) | 63.4 | -28.4 | 16 ± 5 | Modern | |
| *WM 127 m* | | | | | |
| DIC | 829 ± 3 | 0.3 | 45 ± 4 | Modern | OS-82055 |
| POC ( > 0.7 µm) | 5.6 | -29.1 | 12 ± 4 | Modern | OS-82071 |
| Bulk DOC (< 0.7 µm) | 94.4 ± 0.8 | -26.4 | 51 ± 4 | Modern | OS-84604 |
| Init DOC (< 0.2 µm) | 93.3 ± 1.1 | -26.9 | 20 ± 3 | Modern | OS-82425 |
| HMW DOC (>1kDa) | 7.9 | -27.0 | 49 ± 4 | Modern | OS-88679 |
| LMW DOC (< 1kDa) | 85.4 | -26.9 | 17 ± 4 | Modern | |
| *EM 5 m* | | | | | |
| DIC | 827 ± 2 | 0.3 | 39 ± 4 | Modern | OS-82056 |
| POC ( > 0.7 µm) | 4.1 | -29.8 | 33 ± 5 | Modern | OS-82075 |

| | | | | | |
|---|---|---|---|---|---|
| Bulk DOC (< 0.7 µm) | 89.9 ± 1.7 | -26.1 | 45 ± 4 | Modern | OS-82189 |
| Init DOC (< 0.2 µm) | 90.4 ± 0.2 | -27.5 | 47 ± 5 | Modern | OS-82426 |
| HMW DOC (>1kDa) | 10.8 | -26.4 | 60 ± 4 | Modern | OS-88668 |
| LMW DOC (< 1kDa) | 79.6 | -27.7 | 45 ± 5 | Modern | |
| *EM 210 m* | | | | | |
| DIC | 830 ± 3 | 0.4 | 40 + 4 | Modern | OS-82057 |
| POC ( > 0.7 µm) | 4.5 | -29.7 | 48 ± 4 | Modern | OS-82073 |
| Bulk DOC (< 0.7 µm) | 87.9 ± 1.5 | -25.9 | 50 ± 3 | Modern | OS-82190 |
| Init DOC (< 0.2 µm) | 90.4 ± 1.67 | -26.0 | 42 ± 4 | Modern | OS-82502 |
| HMW DOC (>1kDa) | 9.3 | -27.0 | 38 ± 4 | Modern | OS-88669 |
| LMW DOC (< 1kDa) | 81.1 | -25.9 | 42 ± 4 | Modern | |

**Table 4-3. Concentration, stable carbon, and radiocarbon isotopic compositions of size fractionated OM and DIC in thermally stratified Lake Superior in August 2010 and in the Lake Superior tributary Amity Creek (AC) in September 2010.**

| | Concentration (μM) | $\delta^{13}C$ (‰) | $\Delta^{14}C$ (‰) | $^{14}C$ ages (BP yrs) | NOSAMS Accession # |
|---|---|---|---|---|---|
| *AC <1 m* | | | | | |
| DIC | 1869 ± 6 | -11.1 | 42 ± 3 | Modern | OS-84020 |
| POC ( > 0.7 μm) | 24.5 | -29.6 | -127 ± 4 | 1040 ± 35 | OS-84791 |
| Bulk DOC (< 0.7 μm) | 815 ± 8 | -28.3 | 29 ± 4 | Modern | OS-84610 |
| Init DOC (< 0.2 μm) | 776 ± 2 | -28.3 | -4 ± 3 | Modern | OS-84611 |
| HMW DOC (>1kDa) | 21.5 | -28.0 | 36 ± 4 | Modern | OS-84402 |
| LMW DOC (< 1kDa) | 754.5 | -28.3 | -5 ± 4 | Modern | |
| *ONT 4 m* | | | | | |
| DIC | 780 ± 2.0 | 1.1 | 65 ± 5 | Modern | OS-84019 |
| POC ( > 0.7 μm) | 18.9 | -26.4 | 49 ± 3 | Modern | OS-83838 |
| Bulk DOC (< 0.7 μm) | 118.2 ± 1.2 | -26.5 | -8 ± 5 | 40 ± 5 | OS-84603 |
| Init DOC (< 0.2 μm) | 112.4 ± 1.3 | -26.2 | 58 ± 4 | Modern | OS-84818 |
| HMW DOC (>1kDa) | 27.3 | -26.7 | 59 ± 3 | Modern | OS-84379 |
| LMW DOC (< 1kDa) | 85.1 | -26.0 | 58 ± 5 | Modern | |
| *WM 5 m* | | | | | |
| DIC | 780 ± 2 | 1.4 | 57 ± 2 | Modern | OS-83721 |
| POC ( > 0.7 μm) | nd | -28.5 | 53 ± 4 | Modern | OS-83840 |
| Bulk DOC (< 0.7 μm) | 97.3 ± 1.2 | -26.3 | 40 ± 4 | Modern | OS-84142 |
| Init DOC (< 0.2 μm) | 95.7 ± 0.1 | -26.4 | 59 ± 4 | Modern | OS-84655 |
| HMW DOC (>1kDa) | 18.5 | -27.1 | 48 ± 3 | Modern | OS-84399 |
| LMW DOC (< 1kDa) | 77.2 | -26.2 | 62 ± 4 | Modern | |
| *WM 127 m* | | | | | |
| DIC | 806 ± 2 | 0.1 | 56 ± 3 | Modern | OS-83722 |
| POC ( > 0.7 μm) | 4.2 | -29.3 | 12 ± 4 | Modern | OS-83919 |
| Bulk DOC (< 0.7 μm) | 95.9 ± 1.3 | -26.1 | 56 ± 3 | Modern | OS-84212 |

| | | | | | |
|---|---|---|---|---|---|
| Init DOC (< 0.2 μm) | 90.3 ± 1.1 | -26.3 | 53 ± 4 | Modern | OS-84651 |
| HMW DOC (>1kDa) | 12.9 | -26.6 | 60 ± 4 | Modern | OS-84377 |
| LMW DOC (< 1kDa) | 77.4 | -26.3 | 52 ± 4 | Modern | |
| *EM 5 m* | | | | | |
| DIC | 775 ± 4 | 1.5 | 58 ± 3 | Modern | OS-83723 |
| POC ( > 0.7 μm) | 9.3 | -27.3 | 37 ± 4 | Modern | OS-83851 |
| Bulk DOC (< 0.7 μm) | 91.2 ± 0.6 | -26.3 | 2 ± 3 | Modern | OS-84605 |
| Init DOC (< 0.2 μm) | 90.7 ± 1.5 | -26.1 | 44 ± 3 | Modern | OS-84819 |
| HMW DOC (>1kDa) | 11.5 | -27.3 | 57 ± 4 | Modern | OS-84401 |
| LMW DOC (< 1kDa) | 79.2 | -25.9 | 42 ± 4 | Modern | |
| *EM 210 m* | | | | | |
| DIC | 805 ± 2 | 0.2 | 69 ± 3 | Modern | OS-83724 |
| POC ( > 0.7 μm) | nd | -29.2 | -94 ± 5 | 735 ± 40 | OS-83925 |
| Bulk DOC (< 0.7 μm) | 82.9 ± 1.2 | -26.3 | 28 ± 4 | Modern | OS-84606 |
| Init DOC (< 0.2 μm) | 85.5 ± 0.8 | -26.1 | 52 ± 3 | Modern | OS-84820 |
| HMW DOC (>1kDa) | 11.3 | -26.6 | 64 ± 4 | Modern | OS-84400 |
| LMW DOC (< 1kDa) | 74.2 | -26.0 | 50 ± 4 | Modern | |

**Table 4-4. The relative percent distributions of functional groups and compound classes within HMW DOM based on CPMAS $^{13}C$ NMR spectra. The data shows the relative compositions (%) of total carbohydrates, total aliphatic, and total aromatic carbon of HMW DOM from isothermal (in June) and stratified (in August) Lake Superior.**

| Site description and depth | $CH_x$ 0-45 (ppm) | $CH_3$-O 45-60 (ppm) | Total aliphatic | HCOH 60-90 (ppm) | O-C-O 90-120 (ppm) | Total carbohydrate | C=C/Ar-C 120-140 (ppm) | Ar-O 140-160 (ppm) | Total aromatic | Carboxyl COO/CON 160-190 (ppm) | Carbonyl C=O 190-220 (ppm) |
|---|---|---|---|---|---|---|---|---|---|---|---|
| ONT 5m – Isothermal | 14 | 6 | 20 | 48 | 11 | 59 | 2 | 2 | 4 | 15 | 2 |
| ONT 5m – Stratified | 15 | 8 | 23 | 45 | 10 | 55 | 2 | 2 | 4 | 16 | 2 |
| WM 5m – Isothermal | 14 | 6 | 20 | 48 | 12 | 60 | 3 | 2 | 5 | 14 | 1 |
| WM 5m – Stratified | 14 | 7 | 21 | 50 | 11 | 61 | 2 | 2 | 4 | 13 | 1 |
| WM 127m – Stratified | 15 | 9 | 24 | 45 | 10 | 55 | 3 | 2 | 5 | 14 | 2 |
| EM 210m – Isothermal | 14 | 6 | 20 | 47 | 11 | 58 | 2 | 2 | 4 | 17 | 1 |
| EM 5m – Stratified | 15 | 4 | 19 | 58 | 12 | 70 | 1 | 1 | 2 | 9 | 0 |

**Table 4-5. Relative percent compositions of functional groups and compound classes within HMW DOM as obtained from $^1$HNMR.**

| | Maybe organic silicate | Highly branched methyl group | Methyl group of the 6-deoxy sugars | Aliphatic protons of α-carbon attached to carboxylic or aromatic carbon | Acetate ($CH_3CO$) | Aliphatic protons of β, γ carbon attached to carboxylic or aromatic carbon | Total aliphatic | Carbohydrate (HCOH) | Total carbohydrates | Aromatic |
|---|---|---|---|---|---|---|---|---|---|---|
| | 0.50 to 0.20 (ppm) | 0.20 to 0.95 (ppm) | 0.95 to 1.33 (ppm) | 1.33 to 1.76 (ppm) | 1.76 to 2.03 (ppm) | 2.03 to 3.19 (ppm) | | 3.19 to 5.80 (ppm) | | 5.80 to 9.00 (ppm) |
| AC Sept. 2010 | 0 | 9 | 20 | 16 | 9 | 30 | 45 | 7 | 26 | 9 |
| ONT 5m - Isothermal | 3 | 8 | 25 | 12 | 11 | 23 | 35 | 12 | 37 | 5 |
| ONT 5m - Stratified | 1 | 7 | 22 | 13 | 10 | 29 | 42 | 12 | 34 | 7 |
| WM 5m - Isothermal | 2 | 6 | 26 | 14 | 11 | 27 | 41 | 10 | 36 | 3 |
| WM 5m - Stratified | 1 | 8 | 25 | 14 | 11 | 29 | 43 | 8 | 34 | 4 |
| WM 127m - Isothermal | 3 | 7 | 30 | 12 | 13 | 22 | 35 | 11 | 40 | 3 |
| WM 127m - Stratified | 1 | 8 | 25 | 14 | 11 | 30 | 44 | 8 | 33 | 4 |
| EM 5m - Isothermal | 1 | 7 | 24 | 14 | 10 | 32 | 46 | 9 | 33 | 3 |
| EM 5m - Stratified | 3 | 8 | 32 | 10 | 14 | 17 | 27 | 13 | 45 | 3 |
| EM 210m - Isothermal | 3 | 6 | 28 | 12 | 13 | 23 | 35 | 12 | 40 | 3 |

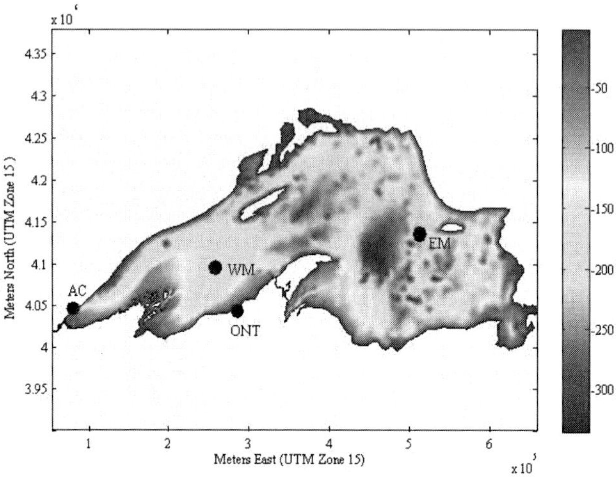


Figure 4-1. Map showing Amity Creek (AC), and nearshore and offshore sites in Lake Superior. The offshore lake sites include western station/mooring (WM) and eastern mooring (EM), and the nearshore site is off Ontonagon River (ONT). The contours are total water depth in meters.

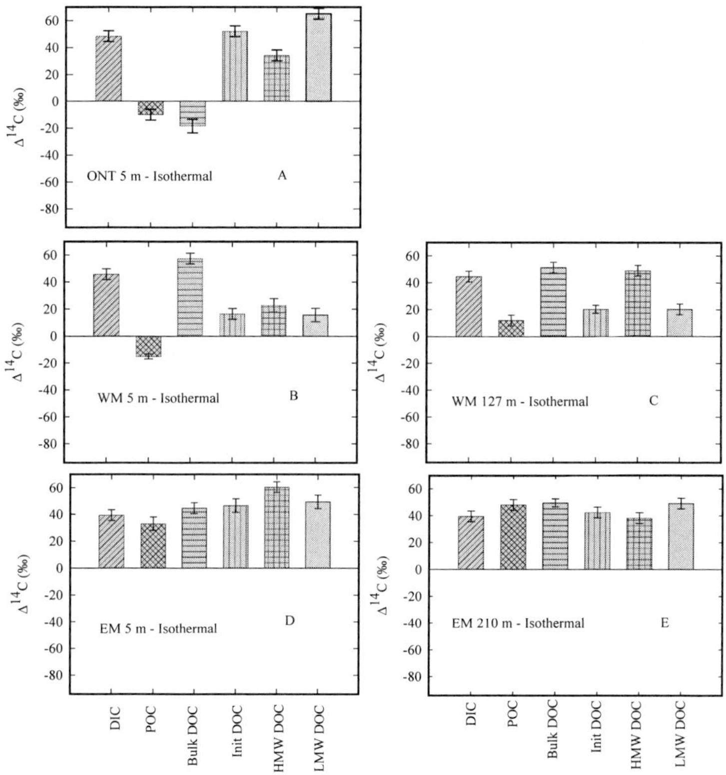


Figure 4-2. The radiocarbon distributions of various organic matter size fractions, and DIC in mixed Lake Superior water column in June.

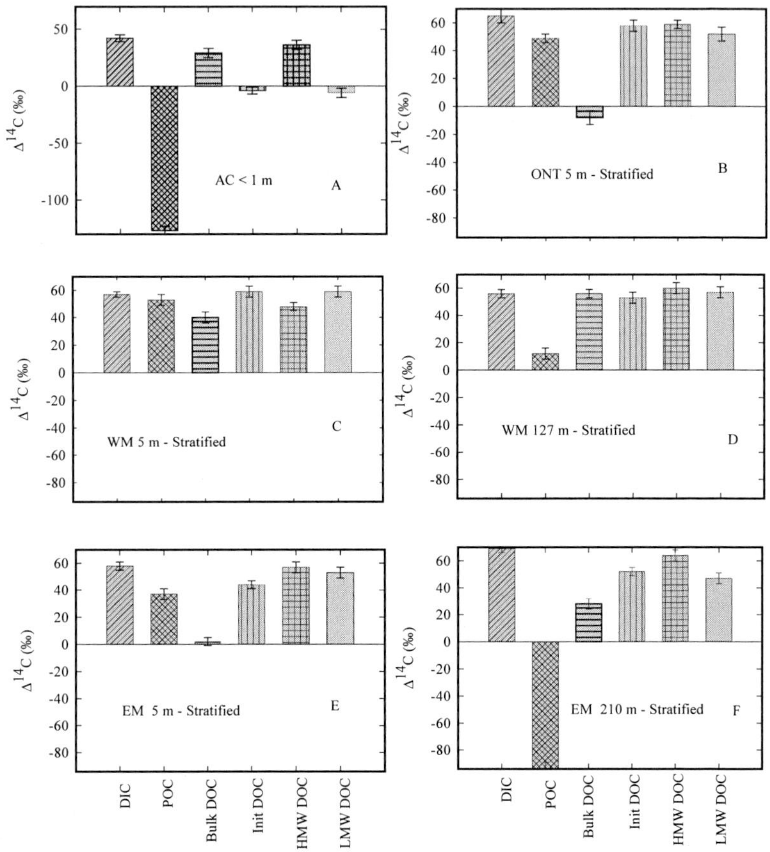


Figure 4-3. The radiocarbon distributions of various organic matter size fractions and DIC in stratified Lake Superior water column in August.

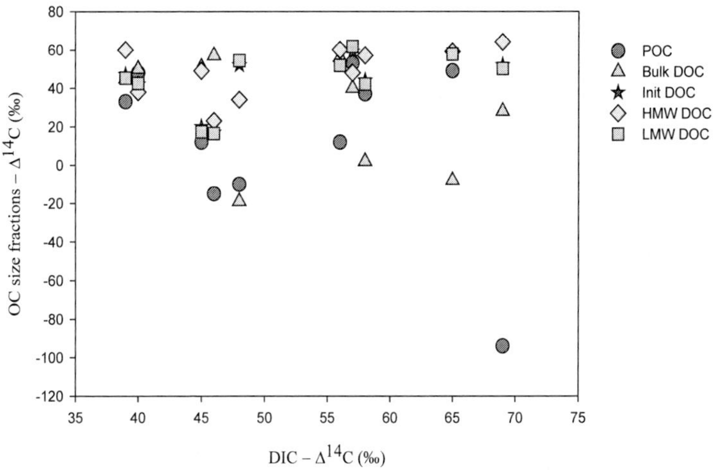


Figure 4-4. Radiocarbon composition of various OC size fractions as a function of $\Delta^{14}C$ of DIC.

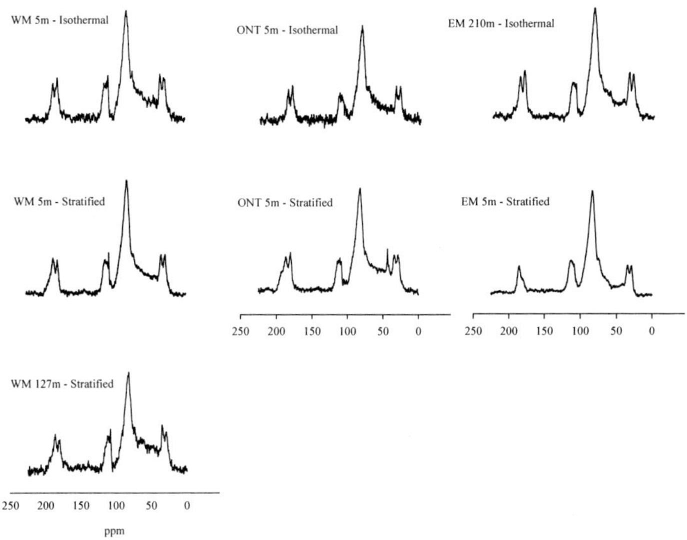


Figure 4-5. The CPMAS $^{13}C$ NMR traces of HMW DOM from different sites and depths in Lake Superior.

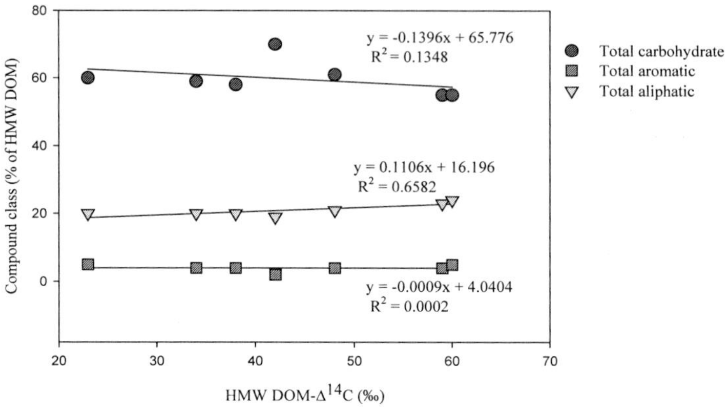


Figure 4-6. The relationship between radiocarbon and compound class composition of HMW DOM.

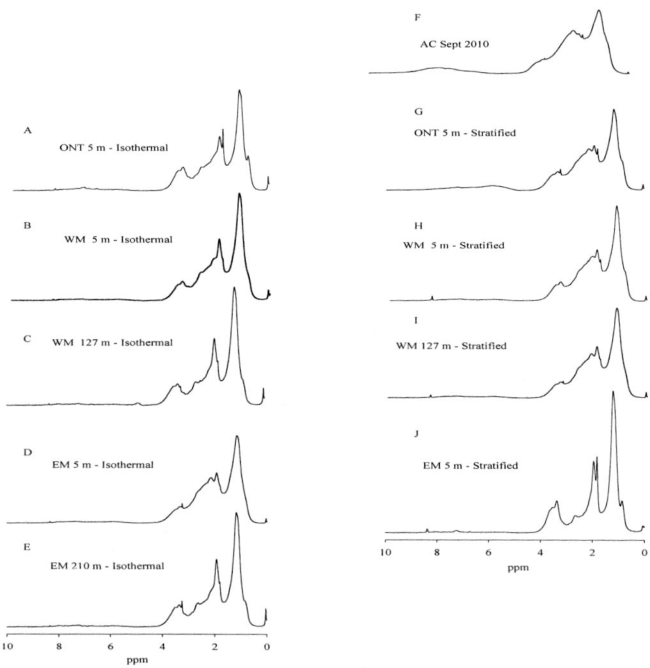


Figure 4-7. $^1$H NMR spectra of HMW DOM from Amity Creek and Lake Superior

# CHAPTER FIVE

# CONSTRAINING THE SOURCES AND CYCLING OF DISSOLVED ORGANIC CARBON IN A LARGE OLIGOTROPHIC LAKE USING RADIOCARBON ANALYSES

## 1.1 Introduction

Dissolved organic carbon (DOC) is ubiquitous in aquatic ecosystems and is important in carbon and energy flow in food webs (Mann et al. 2015; Hitchcock et al. 2016), and in the cycling of nutrients, organic pollutants and trace metals (Riedel et al. 2013). Lakes play important roles in the global carbon cycle with an estimated $CO_2$ emission of 140 Tg C $y^{-1}$ (Cole et al. 1994) and annual carbon burial of up to 58% more than the oceans (Dean and Gorham, 1998). However, the sources and cycling of DOC in freshwater lacustrine systems, especially large lakes, are not well understood. This is primarily because DOC consists of a complex mixture of organic functional groups present in various higher order structures and coming from multiple sources (Hedges 1992; Miller and Zepp 1995). Consequently, DOC has multiple fates. DOC may be routed up to higher organisms in the food web via the microbial loop (Cole et al. 2006; Tanentzap et al. 2014), become sequestered in sedimentary organic matter through particle association (Masiello and Druffel 1998; Coppola et al. 2014) or be mineralized to carbon dioxide via photochemical and microbial oxidations (Obernosterer and Benner 2004).

Constraining the sources and cycling of DOC is a necessary step in understanding functioning of aquatic ecosystems and the global carbon cycle, but the molecular diversity of DOC means that a full separation, identification and characterization of all the organic molecules within it is currently not feasible. These compounds, however, can be grouped into operationally-defined discrete fractions based on their molecular weight (high molecular weight DOC; > 1000 daltons) and polarity (hydrophobic DOC fraction).

In most freshwater systems, a substantial fraction (~30 to 80%) of DOC is comprised of components that can be isolated by adsorption onto a hydrophobic resin (e.g. PPL, C18, XAD-8) at acidic pH, commonly referred to as SPE DOC (Thurman and Malcolm 1981; Dittmar et al. 2008; Sickman et al. 2010). Nuclear magnetic resonance (NMR) analyses of SPE DOC from freshwater lakes and streams show that it is comprised significantly of aliphatic components with substantial contributions attributed to hydrophobic humic substances, aromatic proteins, and lipids/fatty acids (Harvey et al. 1984; Kaiser et al. 2003; Lam et al. 2007; Zhang et al. 2014; Goldberg et al. 2015).

Up to 70% of freshwater DOC, especially in stream, river, or river-impacted coastal systems can be recovered as HMW DOC (nominal molecular weight of 1000 Daltons) using ultrafiltration (Benner and Opsahl 2001; Aluwihare et al. 2002; Stephens and Minor 2010; Zigah et al. 2014). NMR analyses of HMW DOC in large freshwater lakes (and marine systems) show it is rich in complex polymeric structures known as heteropolysaccharides (HPS), peptides/amide and amino sugars (AMS), and carboxylic-rich alicyclic molecules (CRAM) (Hertkorn et al. 2006; Abdulla et al. 2010; Zigah et al. 2014).

In the past few decades, radiocarbon ($^{14}C$) has been useful in studying the ages, cycling and sources of total DOC in freshwater lakes and streams (Trumbore et al. 1992; Alberic et al. 2013; McCallister and del Giorgio 2012; Zigah et al. 2011; Butman et al. 2014; Keaveney et al. 2015). Natural abundance radiocarbon ($\Delta^{14}C$) and stable isotope ($\delta^{13}C$) compositions of the major fractions of DOC provide additional constraints on the multiple sources and cycling of the DOC. $\Delta^{14}C$ calculation corrects for biochemical fractionations so differences in $\Delta^{14}C$ values reflect variable sources and/or cycling. For instance organic material produced from recent within-lake photoautotrophy would bear a $\Delta^{14}C$ value that is similar to that of contemporaneous lake dissolved inorganic carbon (DIC). The few studies in freshwater lakes that do partition DOC show that it consists of a mixture of contemporary and older components (Abbott and Stafford 1996; Sickman et al. 2009; Goldberg et al. 2014). The contemporary DOC components are concentrated in the HMW DOC (Cawley et al. 2016) whereas older DOC components are predominant in the SPE DOC (Abbott and Stafford 1996; Cawley et al. 2016). In a unique closed-lake system, Lake Vida, located in the McMurdo Dry

Valleys of Antarctica that contains primarily microbially-derived DOC with no inputs of DOC from terrigenous sources or recent primary productivity, radiocarbon measurements show that XAD-8 isolated SPE DOC was relatively more $^{14}C$-depleted (older in $^{14}C$ age) than HMW DOC (Cawley et al. 2016). More radiocarbon measurements of DOC fractions from various freshwater systems will help constrain the roles of contemporary and aged carbon in modern biogeochemical cycles and improve our understanding of the global carbon cycle.

This study investigated the isotopic compositions of various organic fractions from Lake Superior, an oligotrophic freshwater lake in North America. We measured the radiocarbon and stable isotope compositions of SPE DOC and HMW DOC from the eastern basin of the lake. We also used serial thermal oxidation and compound class analyses to examine the isotopic diversity of organic matter classes within these two fractions. Together, this data set improves our understanding of the multiple sources and dynamic cycling of DOC in the lake. Lake Superior is a unique and ideal system for this study because the DOC cycling is representative not only of most large freshwater/lacustrine systems, but also the open ocean. Both systems have similar concentrations of DOC (Cotner et al. 2004; Ma and Green 2004), apparent low inputs of allochthonous organic matter and nutrients, a pH range of 7.61 to 8.03 with bicarbonate buffering (Tennant 2016), and a primarily microbial food web (Cotner et al. 2004).

## 5.2 Methods

### 5.2.1 Study site

Lake Superior (Fig. 1) is the Earth's largest freshwater lake by surface area and the deepest of the Laurentian Great Lakes of North America, with a maximum depth of 406 m. The lake is dimictic, with complete vertical mixing of the water column in spring and early winter each year. The hydraulic residence time of the lake (lake volume divided by outflows, including diversions) is 173 years (Quin et al. 1992). DOC is the largest organic carbon pool with an average lakewide surface and deep-water concentration of ~ 90 μM during spring mixing. During summer stratification, lakewide surface DOC is ~ 100-120 μM and is about 10 μM larger than the deep DOC (Zigah et al. 2012). The sources and cycling of DOC in the lake are not well constrained. Annual autochthonous DOC input is estimated at 0.9 Tg C (~ 9% of the primary production; Urban et al. 2005, Sterner 2010), similar to the annual terrestrial DOC loading of

0.4-0.9 Tg C (Cotner et al. 2004; Urban et al. 2005). In spite of the substantial terrestrial input, spectroscopic studies of the DOC, including ultraviolet-visible spectroscopy of total DOC, NMR of HMW DOC, and FTIR of total and HMW DOC have shown fairly low contributions from the aromatic constituents which would indicate terrigenous sources (Minor and Stephens, 2008, Zigah et al. 2014). Hence, the presence, amount and fate of terrigenous DOC in the lake remains poorly understood. To provide greater insights into the composition, cycling and sources of DOC in the lake, we previously investigated the radiocarbon and stable isotope composition of total DOC (Zigah et al. 2011; 2012), and size-fractionated DOC (Zigah et al. 2014). HMW DOC constitutes 8-20% of the total DOC in the western arm and 10-13% of the total DOC in the eastern arm of the lake based on isolation via cross-flow ultrafiltration. NMR spectra of the HMW DOC from the lake show that heteropolysaccharides, peptides/amide and amino sugars together constitute 75-84% of the carbon, with carbohydrate carbon alone representing 53-65% of the HMW DOC (Zigah et al. 2014).

### 5.2.2 Sampling

Surface (5 m) and deep (210 m) samples were collected from eastern Lake Superior (47°34'N, 86°39'W; Fig. 1) during spring mixing in June 2009, and thermal stratification in August 2009 (Fig. 2). Lake water was collected using twelve 8-L Niskin bottles mounted on a rosette equipped with a Seabird Model 911 Plus Conductivity, Temperature, and Depth (CTD) profiler. Dissolved inorganic carbon (DIC) samples were taken directly from the Niskin bottles via pre-cleaned silicone tubing into previously acid-cleaned (10% v/v HCl followed by rinsing with milli-Q water) and combusted (450°C for $\geq$ 4 hours) 0.5 L amber Pyrex bottles. The samples were immediately preserved with 100 µL of saturated mercuric chloride solution, sealed air-tight with glass stoppers coated with Apiezon grease, and stored at room temperature in the dark until analysis. DOC (< 0.7 µm) samples were obtained by filtering lake water through pre-combusted Whatman GF/F glass fiber filters (450°C for 4 hours) using stainless-steel canisters pressurized with nitrogen. Approximately 40 mL of lake water was collected into an acid-cleaned and combusted vial and acidified to pH 2 using 6M HCl for high temperature catalytic oxidation DOC analyses. For DOC radiocarbon analysis, 1 L of the GF/F filtered water was collected into a pre-combusted glass bottle, acidified with 6M HCl to pH 2, and stored in a refrigerator at 4°C until analysis.

### 5.2.3 Ultrafiltration and solid phase extraction

Large-volume samples (200-400 L) (Table 1) were drawn from the lake using an air-driven diaphragm pump, 0.2 μm filtered (Whatman Polycap 75 TC), and ultrafiltered as described previously (Zigah et al. 2014). Previously frozen 0.7 μm-filtered water samples stored in precombusted 1L glass bottles or acid-cleaned 1L high density polyethylene (Nalgene) bottles were thawed, filtered (0.2 μm), and solid phase extracted using the protocol of Dittmar et al. (2008). Briefly, the filtered water was acidified to pH 2 with 6M HCl and extracted using methanol-activated styrene divinylbenzene polymer resin (PPL; 1g, Varian Mega Bond Elut) at a flow rate of 10 mL/ min. The DOC extract was desalted with acidified Milli-Q water (pH 2), dried with ultrahigh purity (UHP) $N_2$ and eluted with methanol. The SPE DOC sample was transferred into quartz combustion tubes and completely dried, amended with precombusted CuO and Ag, evacuated, and sealed on a vacuum line.

### 5.2.4 Measurement of DOC concentrations

DOC and TOC samples were analyzed on a Shimadzu $V_{CSH}$ TOC analyzer (Zigah et al. 2011). Lyophilized HMW DOC samples were analyzed on a Costech ECS 4010 elemental analyzer (EA) coupled to Finnigan Delta Plus XP isotope ratio mass spectrometer (IRMS) after fumigation with 12 N HCl (ACS Plus grade) and drying (Zigah et al. 2014). Typical instrumental precision for OC measurements was 0.2% of the measured concentration. SPE DOC samples were combusted to $CO_2$ in sealed quartz tubes and quantified manometrically. Discrete fractions of $CO_2$ collected during serial thermal oxidation were also quantified manometrically.

### 5.2.5 Extraction of organic fractions from HMW DOC

All frozen HMW DOC samples were freeze-dried and homogenized before analysis. Approximately 75-100 mg of each freeze-dried sample was used for the extractions. The lipid-like, carbohydrate-like and protein-like fractions were extracted from the HMW DOC as described previously (Wang et al. 1998; Loh et al. 2004). The solid residual material after the HCl hydrolysis (unhydrolysable fraction) was also isotopically characterized. In all cases, the extracted organic fractions were transferred into quartz combustion tubes and completely dried, precombusted CuO and Ag were added, and the tubes were evacuated and sealed on a vacuum line.

### 5.2.6 Serial thermal oxidation of HMW DOC and SPE DOC from Lake Superior

Serial thermal oxidation was performed at the National Ocean Sciences Accelerator Mass Spectrometry Facility (NOSAMS) at the Woods Hole Oceanographic Institution (WHOI) using their Ramped PyrOx line (Rosenheim et al. 2008). Dried sample was placed in a reaction vessel which was then placed in the thermal analyzer (two coupled ovens with the bottom oven set at a constant temperature of 800 °C and the top oven holding the sample reactor). With an oxygen (8%) and helium (92%) mixed gas solution flowing through the reactor at a rate of 35 mL per minute, the sample was thermally oxidized to $CO_2$ by gradually raising the top oven temperature from ambient to 1000 °C at a rate of 5 °C per minute. The evolving gas was passed over a Pt/Ni/Cu twisted wire in an 800 °C oven to ensure complete oxidation to $CO_2$. The $CO_2$ was first quantified downstream with an in-line $CO_2$ analyzer, and then collected in successive fractions using flow-through glass traps (Rosenheim et al. 2008; Plante et al. 2013).

### 5.2.7 Radiocarbon and stable isotope measurements

Radiocarbon ($\Delta^{14}C$) and stable carbon isotope ($\delta^{13}C$) measurements were performed at NOSAMS using standard protocols (McNichol et al.1994). Typical instrumental precision of $\delta^{13}C$ based on multiple analyses of standards was 0.15‰. The graphite produced in all cases was analyzed using either a 3MV tandetron accelerator mass spectrometer (USAMS) or 500kV pelletron continuous flow accelerator mass spectrometer (CFAMS) along with primary and secondary standards, and combustion and graphitization process blanks (McNichol et al. 1994). Radiocarbon values are reported as $\Delta^{14}C$ according to the convention of Stuiver and Polach (1977). Instrumental precision of the $\Delta^{14}C$ analysis is based on the error of standards or multiple analyses on a target and was 3-6‰. The radiocarbon composition of lake-water DIC was collated from Zigah 2012, Zigah et al. 2011, 2012 and 2014. The atmospheric $CO_2$ $\Delta^{14}C$ data was based on the radiocarbon composition of corn leaves (*Zea mays*) from the lake watershed (Hseuh et al. 2007; Zigah 2012; Kruger 2014) and from atmospheric $CO_2$ in the Northern Hemisphere (Graven et al. 2012). The atmospheric $CO_2$ data covering 2004-2007 is from Hseuh et al. (2007) and Graven et al. (2012), and the data covering 2009-2012 are from Zigah 2012 and Kruger 2014.

### 5.2.8 Blank assessment

To assess the process blanks and potential fractionation associated with the extractions of organic fractions from HMW DOC for isotopic measurements, we measured the isotopic values of known standard materials before and after extraction. In all cases, the amount of extracted standard material was comparable to or smaller than the extracted samples in terms of $CO_2$ generated after combustion. For the lipid standard (stearic acid), the $\Delta^{14}C$ values of non-extracted and extracted stearic acid were 48 ± 4‰ and 51 ± 4‰, respectively. The $\Delta^{14}C$ values of non-extracted and extracted D-glucose (carbohydrate standard) were 35 ± 4‰ and 41 ± 4‰ and those of bovine serum albumin (protein standard) were 184 ± 3‰ and 179 ± 3‰ respectively. These results indicate that there is a negligible amount of non-modern carbon added during the extraction processes for the organic fractions. Also, non-extracted and extracted $\delta^{13}C$ values of stearic acid (-30.4 ± 0.15‰ vs. -30.4 ± 0.15‰), D-glucose (-11.2 ± 0.15‰ vs. -11.2 ± 0.15‰) and bovine serum albumin (-10.2 ± 0.15‰ vs. -10.6 ± 0.15‰) indicate no fractionation and/or contamination from the extraction processes. The lipid standard we used was modern but the samples we analyzed were pre-aged, so we must evaluate the potential for contamination by modern carbon. If we assume the process blank for the total lipid extraction contains 1 μg C (Loh et al. 2004) with a $\Delta^{14}C$ value of 48‰, similar to the modern stearic acid standard, $\Delta^{14}C$ values of the lipid samples increase by ≤ 1‰. If the process blank contained as much as 12 μg C with a $\Delta^{14}C$ value of 48‰, the lipid samples' $\Delta^{14}C$ values would decrease from the range of -153‰ to -256‰ reported here to -159‰ to -268‰ . This change, while significant, is too small to alter the discussion or conclusions of this paper.

To assess the process blank associated with the SPE DOC extraction, acidified Milli-Q water was extracted using methanol-activated PPL resin, eluted with methanol, dried and combusted to $CO_2$ in sealed quartz tubes at 850°C. Less than 1 μg C was obtained, small relative to the SPE sample size (421-774 μg C), and there was not enough for isotopic analysis. If we assume this carbon has a fraction modern value of 1 ($\Delta^{14}C$ ~ -7‰) or 0 ($\Delta^{14}C$ = -1000‰), the SPE sample $\Delta^{14}C$ values only decrease by ≤ 2‰. To assess the process blank associated with serial thermal oxidation on the Ramped PyrOx (RPO), we cycled the thermal analyzer without any sample in the reactor and quantified the blank carbon to be ~ 4 μg C from ambient to 1000°C, similar to the < 4 μg C reported by Plante et al. (2013). This amount is not enough for isotopic measurement and is very small (≤ 6%) relative to the $CO_2$ amounts in the thermal fractions of the sample

(63-131 μg C). Further evidence of the negligible impact of the process blank on $^{14}C$ results from the RPO comes from the agreement between the $\Delta^{14}C$ values of the bulk HMW and SPE samples and the values calculated for the bulk samples using a mass balance of the RPO data. Due to negligible blank carbon and/or isotopic fractionation, reported $\Delta^{14}C$ and $\delta^{13}C$ values of organic fractions of HMW DOC, SPE DOC and organic fractions from serial thermal oxidation were not corrected for process blanks or fractionation.

## 5.3. Results

### 5.3.1 Trends in the radiocarbon values of Lake DIC and atmospheric $CO_2$ from 2004 to 2012

The $\Delta^{14}C$ value of atmospheric $CO_2$ decreased from ~66‰ in 2004 to 38‰ during our sampling in 2009, and to 25‰ in 2012 (Fig. 3) corresponding to a decline of ~ 5‰/yr. The lake DIC $\Delta^{14}C$ value decreased from 83‰ to 44‰ between 2007 and 2012 with a decline ~ 7‰/yr (Fig. 3). DIC was consistently $^{14}C$-enriched by ~20‰ relative to atmospheric $CO_2$ from 2004-2014, indicating a carbon isotope equilibration time of ~ 4 years between the DIC and atmospheric $CO_2$ over this period.

### 5.3.2 Concentration and isotopic values of HMW DOC vs. SPE DOC vs. Total DOC

The concentration of HMW DOC was 9-11 μM (10-13% of total DOC) with molar C:N values of 14.5-23.6 with little difference between the stratified and mixed lake conditions (Tables 1 and 2). SPE DOC concentrations were 35.5-48.8 μM (41-51% of total DOC) and 60-64 μM (65-71% of total DOC) during stratification and mixed-lake conditions, respectively (Table 2). The $\delta^{13}C$ values of SPE DOC, HMW DOC and total DOC in the lake were -26.4‰ to -26.8‰, -25.9‰ to -26.3‰ and -25.9‰ to -26.3‰, respectively (Table 2). The $\delta^{13}C$ values of total DOC, HMW DOC and SPE DOC did not exhibit considerable variation between spring mixing and summer stratification (Table 2).

The $\Delta^{14}C$ values of SPE DOC (25-43‰) were slightly depleted or similar to the $\Delta^{14}C$ value of atmospheric $CO_2$, and by extension, recent land-plant primary production (38 ± 2‰) from 2009-2010, but more depleted relative to lake DIC ($\Delta^{14}C$ of 59-63‰). The stratified deep and mixed-lake surface

SPE DOC (25-30‰) were more depleted relative to the $\Delta^{14}C$ of lake DIC (Table 2) than were the other two SPE DOC samples. During spring mixing conditions in June, $\Delta^{14}C$ values of HMW DOC were 22-32‰ and increased to 51-54‰ during stratification in August (Table 2). HMW DOC, SPE DOC and total DOC were consistently more depleted in $^{14}C$ than the lake DIC ($\Delta^{14}C$ of 58-59‰ in June and 59-63‰ in August) (Table 2). SPE DOC was more depleted than HMW DOC and total DOC during stratification (Table 2).

### 5.3.3 $\Delta^{14}C$ and $\delta^{13}C$ values of extracted organic fractions within HMW DOC

The solvent extractable lipid-like fraction had $\Delta^{14}C$ values of -204‰ to -256‰ (corresponding to 1770 - 2320 BP years) during spring mixing in June, and -153 ± 13‰ (~1270 BP years) in the stratified surface waters in August (Table 3). The lipid-like fractions were consistently more $^{14}C$-depleted (older) than the concurrent carbohydrate-like ($\Delta^{14}C$ of 20 to31‰) and protein-like ($\Delta^{14}C$ of 34 to 39‰) fractions. The latter fractions were consistently modern, but $^{14}C$-depleted relative to the DIC ($\Delta^{14}C$ of 58 to 63‰) (Tables 2 and 3). The $\delta^{13}C$ values of the carbohydrate-like and protein-like fractions (-25.0‰ to -26.9‰) were more $^{13}C$-enriched than the lipid-like fraction (-28.0‰ to -29.7) (Table 3). The $\Delta^{14}C$ and $\delta^{13}C$ values of the unhydrolysable fraction were variable with depth in the lake (Table 3).

### 5.3.4 Serial thermal oxidation of HMW DOC

The thermogram of the stratified surface HMW DOC shows components oxidizing over the low- temperature range ~ 200-390 °C with two discernable peaks (larger one at ~ 270 °C and smaller one at ~ 380 °C). High-temperature components thermally oxidize at ~ 420-630 °C with peaks at ~513 °C and ~575 °C (Fig. 4a). The thermogram of the mixed-lake surface HMW DOC also shows components oxidizing over low-temperature and high-temperature ranges. A considerably lower percentage of the carbon appeared in the low-temperature range during mixed-lake condition as compared to stratified lake condition (Fig. 4a). The high-temperature peak at ~ 513 °C was smaller in the mixed-lake sample but other high-temperature components were observed including a large peak at ~630 °C (Fig 4a). The thermograms show a strong difference in the composition of thermally refractory material during stratification and mixing conditions. For radiocarbon analysis, six thermal fractions with relative amounts of carbon (in mol %) of 5, 10, 14, 14, 13 and 44 (Fig. 4c) were collected from the mixed-lake surface HMW DOC. The $CO_2$ fractions from the lower temperature range of 200-330 °C were more $^{14}C$- enriched ($\Delta^{14}C$ of 60-64‰) than the $CO_2$ fractions from the higher temperature range of 388-800 °C ($\Delta^{14}C$ of -23 to 34‰) (Fig. 4c). Five thermal fractions with relative amount of carbon (in mol %) of 22, 24, 21, 21 and 12 (Fig. 4e) were collected from the stratified-lake surface HMW DOC. The $CO_2$ fractions from the lower temperature range 200-387 °C were more $^{14}C$-enriched ($\Delta^{14}C$ of 60-75‰) than the $CO_2$ fractions from the higher temperature range 387-800 °C ($\Delta^{14}C$ of 27-30‰) (Fig. 4e). For both mixed-lake and stratified surface HMW DOC samples, the mass-weighted mean $\Delta^{14}C$ value was similar to the independently measured bulk $\Delta^{14}C$ value.

### 5.3.5 Serial thermal oxidation of SPE DOC

The thermogram of the stratified surface SPE DOC shows components oxidizing over a broad temperature range ~ 200-530 °C (peak at 375 °C). High-temperature components, a peak at ~ 504 °C and a shoulder at ~550 °C, were observed in the surface SPE DOC during lake mixing, but these components were absent during stratification (Fig. 4b). Five thermal fractions (F1-F5) with relative amounts of carbon (in mol %) of 13, 17, 17, 20 and 33 (Fig. 4d) were collected from the mixed-lake surface SPE DOC for radiocarbon analysis. SPE DOC components within thermal fractions F1, F4, and F5 were more $^{14}C$-depleted ($\Delta^{14}C$ of -22 to 48‰) relative to the components within fractions F2 and F3 ($\Delta^{14}C$ of 54-58‰) (Fig. 4d). The relative amounts of carbon (in mol %) in the five thermal fractions (F1-F5) from stratified surface SPE DOC were 17, 23, 34, 18 and 8 (Fig. 4f). SPE DOC components within fractions F1, F4, and F5 were more $^{14}C$-depleted ($\Delta^{14}C$ of -8 to 47‰) relative to the components within fractions F2 and F3 ($\Delta^{14}C$ of 60-65‰) (Fig. 4f). The mass-weighted mean $\Delta^{14}C$ values matched the independently measured bulk values of both the mixed-lake and stratified surface SPE DOC.

## 5.4. Discussion

### 5.4.1 Sources and cycling of HMW DOC and SPE DOC fractions

The $\Delta^{14}C$ values of DIC declined from a range of 76‰ to 83‰ in 2007 to a range of 59‰ to 63‰ in 2009 mirroring decreases in the atmosphere.

This implies that DOC originating from planktonic productivity within the lake since the influence of atmospheric bomb $^{14}C$ on the lake DIC (i.e., in the past ~60 years) would bear $\Delta^{14}C$ values ≥ 59‰ (Fig. 3).

During mixed-lake conditions in spring, the radiocarbon values of HMW DOC were more depleted than the values of DIC and recent atmospheric $CO_2$. This suggests there is a significant mobilization of $^{14}C$-depleted (aged) HMW DOC components. The aged DOC must derive from sources other than current primary productivity in the lake, such as the sediments and/or soils (Table 4) or productivity pre-dating the bomb spike in the 1950-60's. The observed $^{14}C$-depletion may be due to a relatively smaller input of autochthonous DOC in the lake during winter and spring when temperature and light levels limit primary production (Sterner 2010), a substantial DOC input from the sediments during lake over-turn, or input from the watershed via spring snowmelt and ice-out (Stottlemyer and Toczydlowski 1991, Zigah et al. 2012). Relative $^{14}C$-enrichment of HMW DOC during stratification is consistent with increased planktonic photosynthesis in the surface waters in the summer months (Urban et al. 2005) or an increase in the contribution from terrestrial productivity (52-60‰) 2-4 years prior to our sampling in the Lake (in 2005-2007) (Fig. 2 and Table 2). However, the dramatically lower C:N values of surface-water HMW DOC in August relative to June (Table 1) coupled with the radiocarbon results suggest an imprint of fresh autochthonous primary production. The radiocarbon values of HMW DOC and total DOC in the stratified deep waters were also enriched, by 19‰ and 16‰ respectively, relative to the values in the mixed-lake (Table 2). This implies rapid vertical transfer within weeks to months of some surface-derived autochthonous DOC to the deep waters. The exact route of DOC delivery to the deep waters is not known, but could be from convective mixing, downwelling events (Austin 2013), and/or solubilization of sinking particles.

Although the isotopic composition shifted significantly, the concentrations of HMW DOC and total DOC only changed slightly (~ 5%) between the mixed-lake and stratified conditions (Table 2). The radiocarbon values of surface HMW DOC and surface total DOC would be 25‰ and 43‰, respectively, in the stratified lake if the additional autochthonous DOC input estimated from concentration changes was the only factor modulating the DOC $\Delta^{14}C$ values. However, the observed radiocarbon values were 54‰ for both HMW DOC and total DOC in the stratified surface waters. The observed changes in radiocarbon can lend insight into the changes in sources and carbon cycling between mixed and stratified conditions. First, it appears that a significant fraction of the $^{14}C$-depleted components of HMW DOC and total DOC in the mixed lake is photochemically and/or microbially remineralized as the lake becomes stratified between June and August. It is possible that the remineralization of CRAM which is a substantial component of HMW DOC is responsible for the observed increase in $\Delta^{14}C$. Zigah et al. (2014) used $^{13}C$ NMR spectra to demonstrate that 16-20% of the HMW DOC from the eastern Lake Superior consists of CRAM, and estimated a value of -230‰ for the $\Delta^{14}C$ of HMW DOC CRAM. Using these values in a mass balance, we estimate that complete oxidation of CRAM will decrease the concentration of HMW DOC by 1.7-2.2 μM and increase the HMW DOC $\Delta^{14}C$ values from 22‰ in the mixed-lake to ~70‰ during stratification. HMW DOC concentration only increased by 0.6-2 μM during stratification, and the $\Delta^{14}C$ value only increased up to 54‰ (Table 2) indicating that not all the CRAM in Lake Superior was oxidized and/or that it is not as $^{14}C$-depleted as we previously estimated. Indeed, the NMR spectra of surface HMW DOC shows the presence of CRAM during stratification at a lower concentration than in the mixed-lake HMW DOC (Zigah et al. 2014).

Second, there may be a considerable mobilization or input of DOC that is more $^{14}C$- enriched in the surface water than the current lake DIC during stratification. For example, autochthonous DOC and terrigenous DOC from the past 5-10 years previous to our sampling in 2009 would have $\Delta^{14}C$ values as high as +120‰; Fig. 3. This is consistent with previous studies in the lake. By extrapolation of the linear relationship between HMW DOC $\Delta^{14}C$ values and HPS and AMS and between HMW DOC $\Delta^{14}C$ values and total carbohydrate carbon from the $^{13}C$ NMR spectra, Zigah et al. (2014) estimated the $\Delta^{14}C$ values of combined HPS and AMS and total carbohydrate carbon within HMW DOC to be 113‰ and 171‰, respectively, which is $^{14}C$-enriched relative to the lake DIC. These results suggest that there is greater routing of carbon through the DOC pool than is evident from the seasonal concentration changes, i.e. the amount of photoautotrophic input during summer stratification is larger than depicted by the change in concentration since there is also removal of some recalcitrant DOC. Removal of $^{14}C$-depleted components such as CRAM coupled with the addition of $^{14}C$-enriched components such as HPS and AMS likely contribute to the isotopic signal of total DOC and HMW DOC in the lake during stratification. This addition and removal of components of HMW DOC is also consistent with the HMW DOC thermochemical fractionation data (section 4.3).

The radiocarbon values of the SPE DOC from the surface water increased from a $\Delta^{14}C$ value of 25‰ in the mixed-lake to 41‰ during stratification while the concentration decreased by 11 μM (18%, Table 2). This suggests that the increase in the $\Delta^{14}C$ could be due to preferential removal of older SPE DOC, possibly through photochemical oxidation. Mass balance calculations indicate that ~11 μM of old SPE DOC with an average $\Delta^{14}C$ value of -44‰ is removed from the surface during thermal stratification. Studies in other lakes have looked at the mechanisms of DOC removal. Photo-irradiation studies have shown that DOC components such as CRAM, black carbon-like molecules and aromatic compounds (e.g. lignin phenols) are highly photo-labile (Spencer et al. 2009; Stubbins et al. 2010, 2012). Cory et al. (2014) noted that photochemical oxidation rather than microbial respiration plays a significant role in DOC mineralization in Arctic freshwater lakes and rivers, accounting for 70-95% of the DOC removal. In a study of 109 Swedish lakes, Kellerman et al. (2015) reported that remineralization processes preferentially remove oxidized and aromatic components of DOC. In their study of oligotrophic high-elevation lakes in California, Goldberg et al. (2015) reported that the total DOC concentration in Lake Tahoe increased from 35 to 44 μM between January and June, but the SPE DOC, isolated with a DAX-8 resin, decreased from 12.3 to 11.9 μM. Also, the SPE DOC was enriched in $^{14}C$ during summer stratification in June ($\Delta^{14}C$ of 35‰) relative to lake mixing in January ($\Delta^{14}C$ of -6‰). A similar enrichment was reported in the SPE DOC from Fallen Leaf Lake in Sierra Nevada during stratification in summer (Goldberg et al. 2015). These observations are consistent with our observations in Lake Superior and suggest that seasonal dynamics in the chemical and isotopic compositions of SPE DOC are controlled by similar processes in these lakes. It appears that $^{14}C$-depleted inputs of SPE DOC during lake mixing and snowmelt in spring and photochemical and microbial removal during stratification in summer are a widespread phenomenon in freshwater lakes.

The relative depletion of $^{14}C$ in a significant fraction of SPE DOC relative to bulk DOC and DIC is consistent with observations by Abbott and Stafford (1995). In a study of three Arctic Lakes, they noted humic substances isolated by XAD-8 resin were depleted in $^{14}C$ relative to total DOC and other carbon pools, but were similar in $^{14}C$ values relative to soils and peat from the watershed. They concluded that these $^{14}C$-depleted terrestrial sources contribute to the lake humic substances. Also, based on multidimensional NMR analysis of SPE DOC from Lake Ontario, Lam et al. (2007) showed that CRAM and materials derived from linear terpenoids potentially constitute the largest component of DOC in the

Laurentian Great Lakes. Terrestrial sources deliver ~0.4-0.9 Tg C to Lake Superior each year (Cotner et al. 2004; Urban et al. 2005), enough to account for the lake's DOC reservoir of 14-17 Tg C (Zigah et al. 2012) and DOC steady state cycling given the hydraulic residence time of 173 years. But evidence for substantial terrigenous DOC in the lake is lacking. Minor et al. (2012) studied the chemical composition of SPE DOC isolated by C18 resin and noted the presence of strong lignin-like, condensed hydrocarbon and aliphatic signatures similar to but modified from terrigenous/riverine SPE DOC that feeds the lake. Seasonal changes in the concentration and isotopic composition of SPE DOC indicate the presence of substantial amounts of $^{14}$C-depleted DOC in the lake.

## 5.4.2 Radiocarbon ages, sources and cycling of organic compound classes within HMW DOC

### 5.4.2.1 HMW DOC Carbohydrate-like and protein-like fractions

Hydrolysable carbohydrate-like and protein-like substances from the HMW DOC were 24‰ to 39‰ depleted in $^{14}$C relative to the DIC, and 7‰ to 18‰ depleted in $^{14}$C relative to the atmospheric $CO_2$ in the sampling year (Table 3, Fig. 5). These indicate that the labile $^{14}$C-enriched carbohydrate-like and protein-like materials from recent in-lake photoautotrophy which have $\Delta^{14}$C values similar to DIC are mineralized rapidly. An estimate of $\Delta^{14}$C of total carbohydrate carbon within HMW DOC based upon the correlation between NMR signals and $\Delta^{14}$C yielded a lakewide carbohydrate $\Delta^{14}$C value of 171‰ (Zigah et al. 2014). The relative depletion of $\Delta^{14}$C values of the hydrolysable carbohydrate-like material observed in this study implies that either the more labile $^{14}$C-enriched components of these hydrolysable substances are not recovered in the hydrolysis, or that there is a significant difference in the carbohydrate structures identified by NMR compared to those recovered by hydrolysis. It is also possible that some $^{14}$C-depleted noncarbohydrate impurities from CRAM or humic substances in the HMW DOC were co-isolated during acid hydrolysis (Repeta et al. 2002). Overall, the modern radiocarbon signatures of the hydrolysable substances in both stratified and mixed-lake conditions (Fig. 5, 6) and the previous NMR-based results showing post-bomb values for the carbohydrate carbon, HPS and AMS constituents indicate that most of these carbohydrate-like and protein-like substances do not persist over very long, century, time scales. The large differences in $^{14}$C-enrichment determined by the two methods used to analyze these fractions of DOC indicate the need for further work linking compound-class characterization and radiocarbon content.

The $\delta^{13}C$ values of DIC in the lake were 0.3-0.4‰ in all but the stratified surface water which had a value of 0.9‰ (Table 2). Aqueous $CO_2$ is typically $^{13}C$-depleted by ~10‰ relative to the bulk DIC (de Kluijver et al. 2014). Applying a typical lake algal photosynthetic fractionation of 17‰ (de Kluijver et al. 2014), organic materials recently synthesized within the lake will have a $\delta^{13}C$ value of -26 to -27‰, similar to the reported algal $\delta^{13}C$ values of -26.0 to -27.8‰ in the lake (Keough et al. 1996). Photochemically altered terrestrially-derived vascular plant sources have similar $\delta^{13}C$ values of -25.0 to -26.6‰ (Vahatalo and Wetzel 2008). Therefore, the $\delta^{13}C$ values of -25.0 to -26.9‰ observed here for the hydrolysable carbohydrate-like and protein-like substances reflect both the values expected from photochemically altered terrigenous sources as well as autochthonous algal-derived sources.

### 5.4.2.2 HMW DOC Lipid-like fraction

Possible sources of the lipid-like fraction are cellular lipids and lipid components of humic substances or petroleum hydrocarbons that may be co-isolated with cellular-derived lipids during solvent extraction (McIntyre et al. 2002). Petroleum hydrocarbons most likely derive from anthropogenic inputs to Lake Superior. The $^{13}C$-depletion of the lipid-like fraction (Fig. 6) is consistent with fractionation associated with cellular lipids biosynthesis (De Niro and Epstein 1977). $\Delta^{14}C$ values of the lipid-like fraction (-256‰ to -153‰) were by far more $^{14}C$-depleted than the acid hydrolysable substances. The millennia ages of the lipid-like fractions are surprising given the hydraulic residence time of the lake (173 years), and suggest that this fraction does not derive from recent bomb $^{14}C$-influenced plankton productivity within the lake or recent terrestrial productivity. In Lake Michigan lipid-like materials from particulate organic carbon (POC) have been reported to be influenced by resuspended sediments (Meyers et al. 1984) and such influences may also occur in the POC and DOC in Lake Superior. The lipid-like fraction and CRAM ($\Delta^{14}C$ of -230‰; Zigah et al. 2014) from HMW DOC are similarly $^{14}C$-depleted suggesting that $^{14}C$-depleted CRAM may be included within the lipid-like fraction. Co-recovery of 70-90% of CRAM with a $\Delta^{14}C$ value of -230‰ or 20-25% of petroleum hydrocarbon with $\Delta^{14}C$ value of -1000‰ along with recently-synthesized cellular lipids would account for the $^{14}C$-depletion observed in the lipid-like fraction relative to the hydrolysable substances. The relative $^{14}C$-enrichment of the surface lipid-like fraction during stratification in August (Fig. 5) could be explained by the addition of ~19% of lake-DIC derived $^{14}C$-enriched 'cellular' lipids from photoautotrophy, or the photochemical removal of ~ 20% of $^{14}C$-depleted components from CRAM. Irrespective of the source, the $^{14}C$-depletion of the HMW DOC lipid-like fraction indicates a $^{14}C$-depleted origin (deep soil horizon or lake sediments or petroleum source; Abbot and Stafford 1995; Sickman et al. 2009; Zigah et al. 2014).

#### 5.4.2.3 HMW DOC Unhydrolysable fraction

The isotopic depletion of the unhydrolysable fraction ($\Delta^{14}C$ = -57‰, $\delta^{13}C$ = -29.5‰; Fig. 6) in deep waters during summer stratification implies that a considerable proportion of this fraction may come from similar sources as the old and $^{13}C$-depleted lipid-like fraction. Selective preservation of lipid-like material has been proposed as the dominant formation mechanism of the unhydrolysable fraction within sinking particulate organic matter in estuaries and oceans (Wang et al. 1998; Hwang et al. 2006). Isotopic enrichment in the surface unhydrolysable fraction during stratification ($\Delta^{14}C$ = 16‰, $\delta^{13}C$ = -27.5‰; Fig 6) suggests some addition from isotopically enriched sources with protein-like and/or carbohydrate-like precursors or removal of some isotopically depleted sources, perhaps through photo-oxidation.

### 5.4.3 Composition and cycling of HMW DOC and SPE DOC based on thermal fractionation

The thermochemical fractionation data show that during spring mixing, about 15% of the surface HMW DOC was in the thermally labile/reactive components, i.e. is oxidized at temperatures < 331°C. These reactive components were $^{14}C$-enriched ($\Delta^{14}C$ values of 60 to 64‰) relative to the more thermally refractory components (Fig 4c). During summer stratification, a significant component (about 46%) of the surface HMW DOC was thermally labile (at <326°C) and $^{14}C$-enriched ($\Delta^{14}C$ values of 70 to 75‰) (Fig 4e). The enriched isotopic values are consistent with an in-lake source reflecting the past 3 years (2007-2009), a recent-past terrigenous source (2004-2005) or a mixture of sources with a substantial post-bomb component (1955-2009) (Fig. 4a). The enriched isotopic values suggest that a substantial component of HMW DOC recycles over shorter times; this is consistent with studies that have reported rapidly recycling semi-labile components within HMW DOC (Amon and Benner 1994; Loh et al. 2004; Repeta and Aluwihare 2006).

In addition to this enriched, thermally labile component, a substantial proportion of the mixed-lake surface HMW DOC (85%) and stratified surface HMW DOC (54%) were more thermally recalcitrant (Fig. 4a, c, e). In the mixed lake HMW sample, the thermally recalcitrant fractions (>331°C) had $\Delta^{14}C$ values ranging from -23 to 34‰, lower than post-bomb atmospheric $\Delta^{14}C$. In the stratified HMW sample, the divide between $\Delta^{14}C$ values higher and lower than the concurrent atmospheric value occurred at a higher temperature range, with fractions >378°C showing $^{14}C$-depletion relative to the atmosphere. The $\Delta^{14}C$ values of the two most thermally stable fractions in each HMW sample were similar to the $\Delta^{14}C$ values of total SPE DOC and HMW DOC in the lake during spring mixing. In a previous study, NMR data showed that HMW DOC consists mainly of two polymeric components - $^{14}C$-enriched HPS and AMS, and older $^{14}C$-depleted CRAM (Zigah et al. 2014). It is likely that the more thermally labile components may include HPS and AMS whereas the thermally recalcitrant components may include the presence of CRAM or humic substances of varying ages. The absence of the large thermally stable components (e.g., peak at ~630 °C) in the surface HMW DOC during stratification is consistent with surface removal by photochemical and/or microbial oxidation of thermally recalcitrant DOC with a lower $\Delta^{14}C$ value. In addition, the relative increase in the thermally labile components during stratification suggests some photoautotrophic inputs during summertime. The stratified-sample enrichment in $^{14}C$ relative to concurrent DIC values for the most thermally labile fractions indicates that there may also be removal of more $^{14}C$-depleted material from these thermally labile fractions. The combination of these processes is consistent with the relative increase in concentration and $^{14}C$-enrichment in HMW DOC during stratification (Table 2).

The data from serial oxidation and acid hydrolysis of HMW DOC differ greatly and indicate that we are isolating different sets of materials with the two techniques; there are several possible mechanisms for this. It is possible that both $^{14}C$-enriched and $^{14}C$-depleted material contribute to the same "compound class" in the acid hydrolysis. These differences could also be explained by differences in recovery for the wet chemical fractions vs. thermal oxidation fractions. Also, isotopic mixing during thermal fractionation could lead to the observed disparity. For instance, the depleted lipid-like fraction could be easily mixed into one of the thermal fractions due to their small relative abundance. Coupling both acid hydrolysis and thermal oxidation appears to be better in constraining the isotopic heterogeneity, and by extension, the sources and cycling of HMW DOC in the lake. However, better distinguishing the windows of view into the DOC pool that each approach provides is an area needing further study.

The thermograms of SPE DOC show that the thermally-labile components in the mixed and stratified samples overlap considerably. The $^{14}C$ values for the thermal fractions show that both old and contemporary material contributes to the SPE DOC (Fig. 4d). About 57% of the stratified surface SPE DOC (mean $\Delta^{14}C$ values of 59-65‰) and 34% of the mixed-lake SPE DOC (mean $\Delta^{14}C$ values of 54-58‰) was $^{14}C$-enriched, similar to the lake DIC and terrestrial productivity 2-4 years prior to our sampling in the Lake (in 2005-2007), and derived from contemporary sources, possibly hydrophobic/aromatic proteins (Goldberg et al. 2014). Also, substantial components of the SPE DOC in the mixed-lake (66% with $\Delta^{14}C$ values of -22 to 48‰) and stratified lake (43% with $\Delta^{14}C$ values of -8 to 47‰) were $^{14}C$-depleted relative to the lake DIC. The most thermally refractory portion, with pre-bomb $^{14}C$ values, was considerably smaller during stratification (Fig. 4 b, d, and f). However, the most thermally labile SPE DOC component in each SPE sample was also $^{14}C$-depleted, especially in the stratified sample (Fig 4d, f). As with the HMW DOC, thermally recalcitrant components present in the surface SPE DOC during lake mixing in spring appear to be removed from the surface SPE DOC during stratification consistent with summertime removal of thermally recalcitrant and old DOC from the lake (Fig 4b, d, f). In contrast to HMW DOC, there does not appear to be a considerable input of recently synthesized organic material into the thermally labile portion of the stratified lake SPE sample, and, in fact, the most thermally labile fraction actually shifted to pre-bomb values (Fig 4b, f). Overall, these observations suggest there was little contribution to the SPE DOC from photoautotrophy during summertime, and are consistent with the decrease in concentration and relative $^{14}C$-enrichment observed in surface SPE DOC during stratification (Table 2).

## 5.5. Summary and conclusions

Our isotopic analyses teased apart the relative influence of contemporary and older DOC in the lake ecosystem. We show that there is a significant mobilization of $^{14}C$-depleted DOC in the water column during spring mixing. Thermochemical fractionation data indicates that this older DOC likely derives predominantly from thermally refractory components. Summer lake stratification enhanced the removal of the older recalcitrant DOC in the lake, implying strong roles for photochemical and microbial remineralization in the sunlight-filled, warmer surface-layer of the lake.

Climate change induced effects on the extent and duration of ice cover and summer stratification could impact DOC cycling, nutrient availability and carbon and energy transfers in the lake food web in the future.

**Table 5-1. Sampling depth, sample volumes, and elemental composition of HMW DOC isolated from eastern Lake Superior in June and August 2009. Total water depth at this site is 242 m.**

| Sample | Depth (m) | Total volume (L) | Retentate volume (L) | Molar C:N |
|---|---|---|---|---|
| *Mixed - June 2009* | | | | |
| Surface | 5 | 200 | 1.25 | 23.6 |
| Deep | 210 | 400 | 0.90 | 17.7 |
| *Stratified - August 2009* | | | | |
| Surface | 5 | 400 | 1.45 | 14.5 |
| Deep | 210 | 383 | 1.20 | 17.8 |

**Table 5-2. Concentrations and carbon isotopic compositions of total DOC (< ~0.7 µm), high molecular weight DOC (HMW DOC, > 1kDa, <0.2 µm), solid phase extracted DOC (SPE DOC, <0.2 µm) and dissolved inorganic carbon (DIC) from eastern Lake Superior during isothermal condition in June and thermal stratification in August 2009.**

| Season / Depth | DIC | | | Total DOC | | | HMW DOC | | | SPE DOC | | |
|---|---|---|---|---|---|---|---|---|---|---|---|---|
| | µM | $\delta^{13}C$ (‰) | $\Delta^{14}C$ (‰) | µM | $\delta^{13}C$ (‰) | $\Delta^{14}C$ (‰) | µM | $\delta^{13}C$ (‰) | $\Delta^{14}C$ (‰) | µM | $\delta^{13}C$ (‰) | $\Delta^{14}C$ (‰) |
| Mixed – June 2009 | | | | | | | | | | | | |
| 5 m | 819 | 0.4 | 59 ± 4 | 91.7 | -26.3 | 42 ± 4 | 10.8 | -26.1 | 22 ± 4 | 60.0 | -26.8 | 25 ± 3 |
| 210 m | 830 | 0.3 | 58 ± 2 | 90.4 | -26.0 | 30 ± 4 | 9.3 | -25.9 | 32 ± 3 | 64.0 | -26.7 | 43 ± 2 |
| Stratified – August 2009 | | | | | | | | | | | | |
| 5 m | 810 | 0.9 | 59 ± 4 | 95.7 | -26.0 | 54 ± 3 | 11.4 | -26.0 | 54 ± 3 | 48.8 | -26.5 | 41 ± 3 |
| 210 m | 824 | 0.3 | 63 ± 4 | 85.5 | -25.9 | 46 ± 4 | 11.3 | -26.3 | 51 ± 5 | 35.5 | -26.4 | 30 ± 4 |

**Table 5-3. Radiocarbon and stable carbon isotopic composition of organic fractions extracted from HMW DOC from eastern Lake Superior in 2009. Instrumental precision of $\Delta^{14}C$ and $\delta^{13}C$ analyses are 3-5‰ and 0.15‰, respectively. The values in parenthesis are the ages in years BP.**

| | Mixed-lake | | Stratified-lake | |
|---|---|---|---|---|
| Organic fraction | $\delta^{13}C$ (‰) | $\Delta^{14}C$ (‰) | $\delta^{13}C$ (‰) | $\Delta^{14}C$ (‰) |
| **Surface (5 m)** | | | | |
| Lipid-like | -28.0 | -204 (1770) | -29.1 | -153 (1270) |
| Carbohydrate-like | -25.8 | 31 | -26.9 | 20 |
| Protein-like | -25.0 | nd | -26.2 | 39 |
| Unhydrolysable material | -26.6 | 59 | -27.5 | 16 |
| **Deep (210 m)** | | | | |
| Lipid-like | -29.7 | -256 (2320) | nd | nd |
| Carbohydrate-like | -26.9 | 28 | nd | nd |
| Protein-like | -25.2 | 34 | -25.7 | 36 |
| Unhydrolysable material | -26.7 | 25 | -29.5 | -57 |

nd = sample lost during extraction, combustion to $CO_2$ or graphitization.

**Table 5-4. Elemental and isotopic composition of putative sources of HMW DOC in the water column of Lake Superior.**

| Putative sources | C:N | $\delta^{13}C$ (‰) | $\Delta^{14}C$ (‰) | Reference |
|---|---|---|---|---|
| In situ primary production | 8-16 | -26 to -27.8 | 61 | Keough et al. 1996; this study |
| Recent terrestrial vascular plant | 35-48 | -25 to -30 | 38 | Lara et al 1998; Lobbes et al. 2000; Guo et al, 2003; this study |
| Terrestrial vascular plants ca. 30-50 yr | 35-48 | -25 to -30 | 200-500 | Lara et al 1998; Lobbes et al. 2000; Guo et al, 2003; this study |
| Lake sediment organic carbon | 10-12 | -25.6 to -27.9 | -36 to -180 | Zigah et al. 2012; Li et al 2013 |
| Surficial soil organic carbon | 10-12 | -26 to -28 | ca. 50 to -50 | Peterson and Fry 1987; Fry 1991; Marwick et al. 2015 |
| Deep horizon soil organic carbon | 10-12 | -26 to -28 | ca. -200 to -400 | Peterson and Fry 1987; Fry 1991; Marwick et al. 2015 |

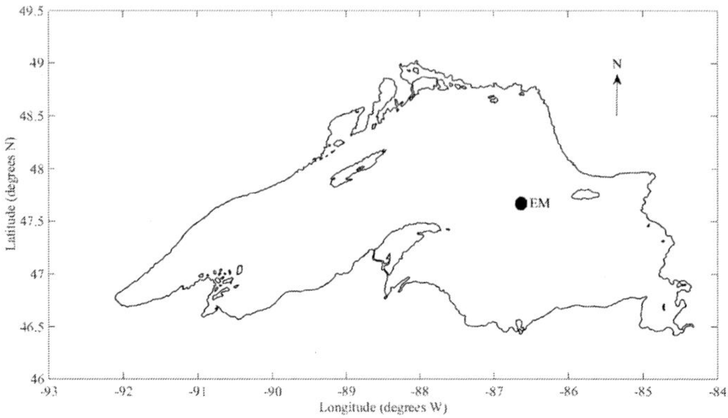


Figure 1. Sampling site in the eastern basin of Lake Superior. EM means eastern mooring.

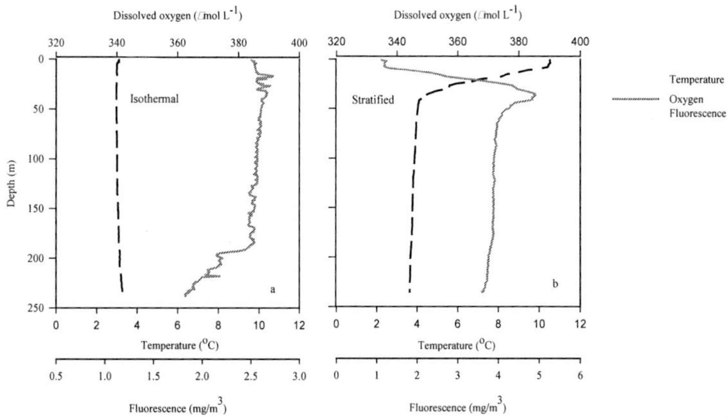


Figure 2. Depth profiles of (a) dissolved oxygen, temperature and chlorophyll-a florescence in the isothermal water column of Lake Superior in June 2009, and (b) dissolved oxygen, temperature and chlorophyll-a florescence (WETLabs WETStar, mg/m$^3$) in the stratified water column of Lake Superior in August 2009. The oxycline and the thermocline were observed at ~2-40 m, and the entire column remained oxygenated during stratification in August.

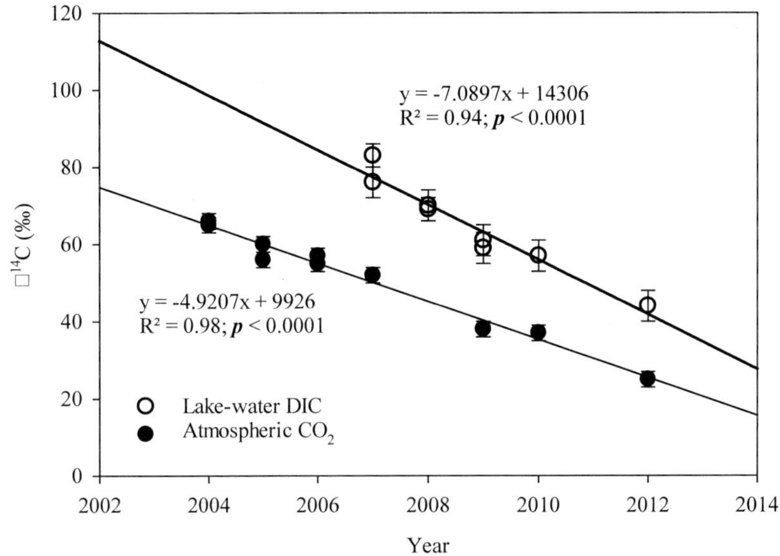


Figure 3. Radiocarbon compositions of lake-water DIC from surface (5 m) and deep chlorophyll maximum (30-40m) depths and atmospheric $CO_2$ covering the years 2004-2012. The $\Delta^{14}C$-DIC data was collated from Zigah 2012 and Zigah et al. 2011, 2012 and 2014. Where available, both the surface (5 m) and depth of deep chlorophyll maximum (30-40 m) were included. Data from 2009-2010 were from the eastern region of the lake (and were similar to those from the western region of the lake) and data from 2007, 2008, and 2012 were from the western region of the lake (which we assume to be same as those at the eastern region of the lake given the similarity of the 2009-2010 data in particular, and homogeneity of $\Delta^{14}C$-DIC in the lake in general as reported by Zigah et al. 2012). The atmospheric $CO_2$ $\Delta^{14}C$ data was adapted from Zigah 2012, Hseuh et al. 2007, Graven et al. 2012 and Kruger 2014.

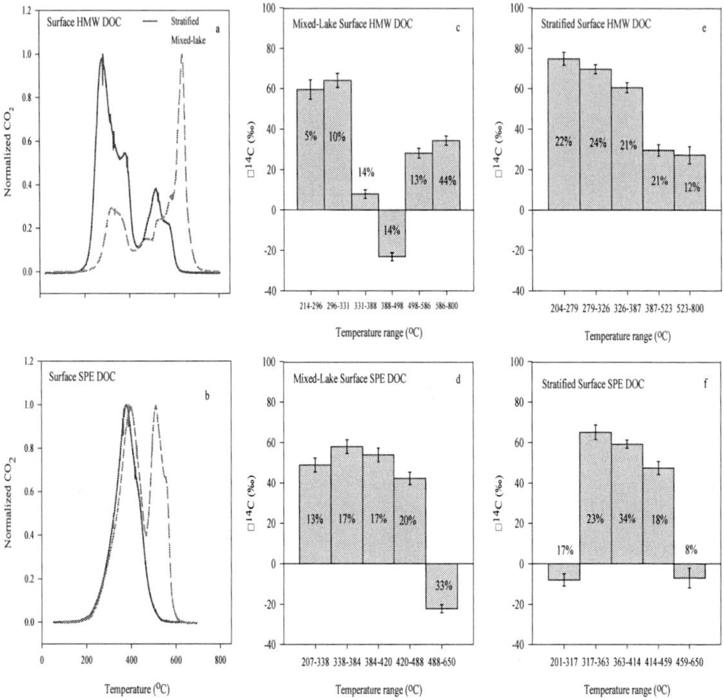


Figure 4. Thermogram of (a) surface HMW DOC during stratification and mixed-lake conditions and (b) surface SPE DOC during stratification and mixed-lake conditions from eastern Lake Superior. $CO_2$ values were normalized to the height of base peak. Different components were thermally oxidized at different temperature ranges. The radiocarbon values of the isolated thermal fractions are given for surface HMW DOC during mixed-lake (c) and stratified conditions (e) for surface SPE DOC during mixed lake (d) and stratified conditions (f). The values within the histograms are the carbon mole percent of each thermal fraction. The error bars are $1\sigma$ accelerator mass spectrometer (AMS) instrumental precision.

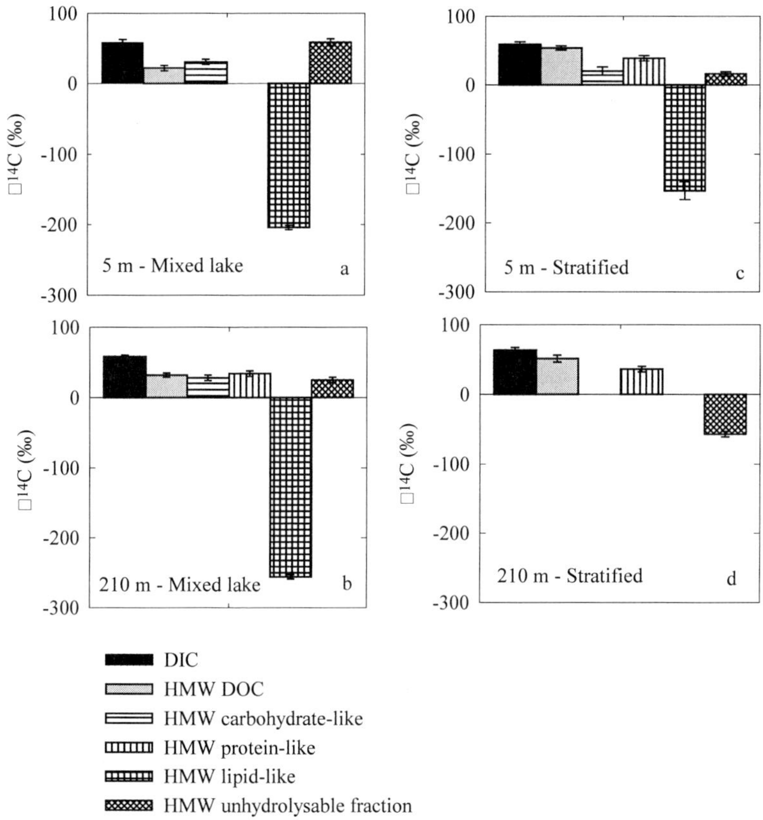


Figure 5. Radiocarbon composition of DIC, SPE DOC, HMW DOC, and compound classes isolated from HMW DOC (carbohydrate-like, protein-like, lipid-like and unhydrolysable fractions from (a) the surface waters (5 m) during mixed-lake condition (b) the deep waters (210 m) during mixed-lake condition (c) surface waters during stratification (d) deep waters during stratification.

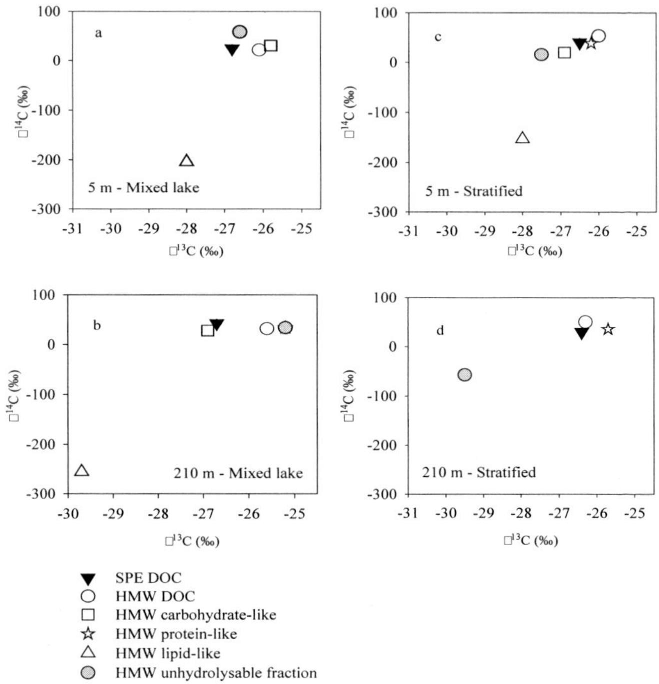


Figure 6. $\Delta^{14}C$ and $\delta^{13}C$ cross-plots of SPE DOC, HMW DOC and organic fractions from HMW DOC from (a) the surface waters (5 m) during mixed-lake condition (b) the deep waters (210 m) during mixed-lake condition (c) surface waters during stratification (d) deep waters during stratification. The lipid-like fraction was more depleted than the other organic fractions.

# CHAPTER SIX

# AN ISOTOPIC INVESTIGATION OF THE COMPOSITION OF PARTICULATE ORGANIC MATTER AND ZOOPLANKTON FOOD SOURCES IN LAKE SUPERIOR AND ACROSS A SIZE-GRADIENT OF AQUATIC SYSTEMS

## 6.1 Introduction

The role of terrigenous organic carbon in aquatic food webs is not yet well constrained. Some studies (Pace et al., 2004; Carpenter et al., 2005; Cole et al., 2006) have reported that terrigenous organic carbon supports disproportionately high zooplankton production in lakes. Others have suggested a smaller role for allochthony (the consumption of organic matter produced outside of the system of interest), and have noted that, rather, phytoplankton sustain disproportionately larger and/or most of the zooplankton production in aquatic ecosystems (Brett et al., 2009; Karlsson, 2007; Pace et al., 2007; Karlsson et al., 2012). Still others have reported seasonal shifts in the food resources supporting aquatic food webs such that autochthony (consumption of organic matter produced within the system of interest) is predominant during high within-lake phytoplankton productivity in summer whereas allochthony (as well as heterotrophic bacterial biomass) is most important to zooplankton biomass during winter periods when within-lake primary and secondary production is minimal (Grey et al., 2001; Taipale et al., 2008; Karlsson and Sawstrom, 2009; Rautio et al., 2011).

There has been a realization that terrigenous organic matter exported from catchments is less refractory within aquatic systems than previously recognized, and can fuel microbial metabolism (Jones and Salonen, 1985; Tranvik, 1992), and that even the ancient (old according to radiocarbon measurements) component traditionally thought to be more recalcitrant could support bacterial (Cherrier et al., 1999; Petsch et al., 2001;

McCallister et al., 2004), zooplankton (Caraco et al., 2010), and fish production (Schell, 1983). Accordingly, the notion that terrestrial carbon partially sustains food webs in aquatic systems has gained currency in the past few decades (Salonen and Hammar, 1986; Hessen et al., 1990; Meili et al., 1993; Pulido-Villena et al., 2005; Cole et al., 2011). Terrigenous carbon could be introduced and accumulated in aquatic food webs by zooplankton directly feeding on terrestrially-derived detrital particles (Hessen et al., 1990; Cole et al., 2006; Brett et al., 2009), and/or feeding on heterotrophic organisms that consume terrestrially-derived organic carbon (Jones, 1992; Lennon and Pfaff, 2005; Berggren et al., 2010).

Several studies in lakes have concluded that terrigenous food can support aquatic animal consumers (Cole et al., 2006; Karlsson and Sawstrom, 2009), and the relative importance of allochthony in lakes is thought to relate to factors such as lake color (indicating the amount of humic material present), trophic status, and size. Therefore, allochthony should be higher in small humic lakes, and lower in eutrophic lakes and/or clear-water lakes with less terrestrial influence on organic matter cycling (Jones, 1992; Pace et al., 2007; Cole et al., 2011). The relative significance of these factors has been difficult to test as neither lab-based studies (Salonen and Hammar, 1986; Brett et al., 2009), small-scale in situ enclosure studies (Hessen et al., 1990) nor whole-lake $^{13}C$-labeled bicarbonate addition approaches (Cole et al., 2002; Carpenter et al., 2005; Cole et al., 2006; Pace et al., 2007; Taipale et al., 2008) are easily applied to large-lake or marine systems. Also, the use of whole-lake $^{13}C$ labeling techniques for estimating the proportion of terrigenous organic carbon supporting zooplankton in lakes is limited or challenged by the fact that unlabeled food particles incorporated by zooplankton could possibly be from metalimnetic phytoplankton or phytoplankton-derived material predating label introduction rather than from terrestrial sources (Brett et al., 2009). Further, quantification of zooplankton food sources using ambient stable carbon isotopic signatures is difficult because of the inherent difficulty in directly measuring the $\delta^{13}C$ of phytoplankton, and the narrow and overlapping range of phytoplankton and terrigenous organic matter $\delta^{13}C$ signatures, especially in freshwater systems (Hamilton et al., 2005). The dynamic range of $\Delta^{14}C$ (-1000 to ~+200‰) is much greater than that of $\delta^{13}C$ in organic carbon (-32 to -12‰) (Petsch et al., 2001; McCallister et al., 2004; Wakeham et al., 2006), and provides a more sensitive means for differentiating the sources of organic carbon in the particulate organic matter (POM) matrix and organic carbon sustaining zooplankton secondary production. Also, while both $\delta^{13}C$ and $\Delta^{14}C$ are linear quantities that can be used for isotopic mixing models, $\Delta^{14}C$ has the added advantage of being the same for consumers and their food source in a modern ecosystem (as the $\Delta^{14}C$ calculation corrects for biochemical fractionations) thereby eliminating the need for fractionation correction along trophic levels as is the case for $\delta^{13}C$ (and $\delta^{15}N$).

In this study we examine the possible food sources of mesozooplankton in Lake Superior, the world's largest freshwater lake by surface area (Herdendorf, 1990), using natural abundance radiocarbon distributions. Recent investigations of Lake Superior, an oligotrophic system with low nutrient concentrations and primary productivity and a pronounced deep-chlorophyll maximum (Russ et al., 2004; Barbiero and Tuckman, 2004), have concluded that the lake appears to be net heterotrophic (McManus et al., 2003; Cotner et al., 2004; Russ et al., 2004; Urban et al., 2004; Urban et al., 2005). Terrigenous and resuspended sedimentary organic carbon sources have radiocarbon signatures that are unique and different from those of the lake's dissolved inorganic carbon and recently fixed primary production, hence providing the opportunity for better understanding the role of these possible food sources in mesozooplankton production and food web dynamics in the lake.

We exploit the natural abundance of radiocarbon ($\Delta^{14}C$), stable isotope ($\delta^{13}C$ and $\delta^{15}N$), and elemental compositions (atomic C:N ratio) of mesozooplankton to assess the role of different carbon sources in supporting mesozooplankton production, thereby providing a clearer picture of food web dynamics in Lake Superior. We also assess the putative food sources of zooplankton in a suite of other aquatic systems (riverine, smaller-lakes, and oceanic) for a broader-scale understanding of zooplankton food sources in aquatic food webs.

## 6.2. Methods

### 6.2.1 Sampling

Cruises were undertaken on the R/V *Blue Heron* to sample Lake Superior in May-June and August-September 2009 during isothermal (mixed) and thermally stratified water conditions, respectively. Site locations, water depths, and sampling depths are given in Fig 1 and Table 1. At each of the eight (8) sampling sites, we first obtained temperature, chlorophyll and depth profiles using a Seabird model 911 plus conductivity, temperature, and depth (CTD) rosette equipped with fluorometer, dissolved oxygen sensor, colored dissolved organic matter (CDOM) sensor, and altimeter (See supplementary Fig 1 A-L). For each of our open-lake stations, where water-column depths ranged from 155 to 388 m, we collected mesozooplankton using 50 m vertical tows through the water column using a 300 $\mu$m plankton net. At each of the nearshore stations (ONT and BR) the depth of tow was modified to a maximum depth of 4 to 10 m above the sediment water interface. The biomass was rinsed with lake water into the cod end of the net and duplicate samples were filtered onto glass-fiber filters (precombusted GF/F filters, 0.7 $\mu$m pore size), and stored frozen. Although we did not separate mesozooplankton into different groups in this study, a recent survey in the lake shows that copepods are the most dominant zooplankton in the surface waters of offshore Lake Superior (Yurista et al., 2009). In this extensive study at 31 sites over a 3 year period, Yurista et al. (2009) reported ~90% (by biomass) of the crustacean zooplankton in the offshore sites (>100 m water depth region) were copepods, and most of these (~80%) were concentrated in the surface 50 m of the lake water column, which is the depth over which we sampled our zooplankton in the offshore lake. Within the copepods, the taxa calanoids dominated by *Diaptomus copepodites* and *Limnocalanus macrurus* were more abundant in the lake than the cyclopoids which were mostly *Cyclops bicuspidatus thomasi* and *Cyclops copepodites* (Yurista et al., 2009). The calanoid copepods contributed ca. 70% of the biomass of crustacean zooplankton in Lake Superior (State of the great lakes report, 2009). For the copepods, the average volumetric concentration by biomass (and by numbers) of adult calanoids, immature calanoids, adult cyclopoids, and immature cyclopoids in net samples were 6255 µg $m^{-3}$ (205 $m^{-3}$), 9682 µg $m^{-3}$ (1864 $m^{-3}$), 1197 µg $m^{-3}$ (236 $m^{-3}$), and 1305 µg $m^{-3}$ (864 $m^{-3}$), respectively (Yurista et al., 2009). For cladocerans, the concentration of bosminids and daphnia were 633 µg $m^{-3}$ (342 $m^{-3}$) and 4203 µg $m^{-3}$ (309 $m^{-3}$), respectively (Yurista et al., 2009).

Sediment cores were taken from the open lake sites using an Ocean Instruments multi-corer. Recovered cores were sectioned at 2 cm resolution and kept frozen until further analysis, and the surface sediments (top 0-2 cm inclusive of the flocculant layer) were used in this study.

We collected dissolved inorganic carbon (DIC), dissolved organic carbon (DOC), and particulate organic carbon (POC) samples from surface waters (≤5 m water depth). Water samples were drawn using twelve 8 L Niskin bottles mounted on the CTD rosette. DIC samples were collected directly from the Niskin bottles via pre-cleaned (10% HCl v/v, then ultra pure water [Millipore Milli-Q Plus]) silicone tubing into previously acid-cleaned and combusted (450°C for 4 hours) 500 mL amber Pyrex bottles.

The bottles for DIC were rinsed three times with sample and then overflowed with two volumes of the unfiltered water. As quickly as possible after collection a small aliquot of water was removed, and the samples were preserved with saturated mercuric chloride solution, sealed airtight with glass stoppers coated with Apiezon M grease and stored at room temperature in the dark. POC and DOC samples were obtained by filtering lake water through precombusted Whatman GF/F glass fiber filters (450°C for 4 hours; 0.7 μm nominal pore size) via nitrogen pressurized stainless-steel canisters. Approximately 1 L of DOC sample from the resulting filtrate was collected into an acid cleaned and combusted glass bottle, acidified to pH 2 using 6N HCl (American Chemical Society Plus grade) and refrigerated. After ~10 L of lake water had passed through a GF/F filter, the filter with retained particulate matter (POC sample) was placed in previously-combusted aluminum foil and stored frozen until analysis.

For comparison with our Lake Superior study we collated similar data for a suite of aquatic ecosystems of various sizes and residence times. Data from five northern small lakes sampled between June and September 2004 in Southern Quebec were adapted from McCallister and del Giorgio (2008); these small lakes include Bran-de-Scie, Des Monts, Stukely, Bowker, and Fraser Lakes. Zooplankton biomass and water samples for DOC, DIC, and POC and their isotopic signature were collected at a depth of 0.5 – 1.0 m using a diaphragm pump connected to an acid rinsed (10% HCl) plastic hose (McCallister and del Giorgio, 2008). Zooplankton were collected by passing at least 200 L of water through a 50 $\mu$m mesh screen, subsequently washed from the screen and stored overnight in deionized water at 4 °C to evacuate gut contents prior to isotopic analysis (McCallister and del Giorgio, 2008). The zooplankton samples were dominated by cladocerans and copepods. Cladocerans were primarily comprised of the genus Daphnia, most notably by *Daphnia mendotae* and *Daphnia catawba*, while copepods were dominated by *Diacyclopsbicuspidatus*, *Mesocyclops edax*, and *Letptodiaptomus minutus*. We also collated existing data from the Pacific Ocean (including Pacific coastal ocean, North Central Pacific, and North Eastern Pacific sites) and the Hudson River (eastern New York, USA). Data from the Hudson River was adapted from Caraco et al. (2010). Pacific Ocean zooplankton data included crustaceans and fishes, and were adapted from William et al. (1987), Druffel and William (1990), Druffel and William (1991), and Druffel et al. (1996).

### 6.2.2 Radiocarbon ($\Delta^{14}C$) analysis

$\Delta^{14}C$ measurements for Lake Superior samples were performed at the National Ocean Sciences Accelerator Mass Spectrometry Facility (NOSAMS) at the Woods Hole Oceanographic Institution (WHOI). POC, zooplankton, and sediments were freeze-dried, acid fumigated (12 N HCl) overnight, and redried, and then converted to $CO_2$ by combustion in a modified Carlo Erba NA 1500 elemental analyzer (Zigah et al., 2011). DOC and DIC samples were converted to $CO_2$ by ultraviolet irradiation and phosphoric acid volatilization, respectively. The evolved $CO_2$ was cryogenically separated and reduced to graphite with $H_2$ over Fe catalyst (Zigah et al., 2011). The graphite produced was analyzed by accelerator mass spectrometry (AMS) along with primary and secondary standards, and combustion and graphitization process blanks.

Radiocarbon values are reported as $\Delta^{14}C$, the part per thousand deviation of the sample's $^{14}C$:$^{12}C$ ratio relative to a nineteenth century wood standard that has been corrected to the activity it would have had in 1950 and a $\delta^{13}C$ of -25‰. $\Delta^{14}C$ was corrected for fractionation using $\delta^{13}C$ of samples according to the convention of Stuiver and Polach (1977). Instrumental precision of the $\Delta^{14}C$ analysis is based on the error of standards or multiple analyses on a target.

### 6.2.3 Stable isotopes ($\delta^{13}C$ and $\delta^{15}N$) and C:N analysis

Carbon and nitrogen contents of bulk POM and zooplankton were measured on a Costech ECS 4010 elemental analyzer (EA) coupled to a Finnigan Delta Plus XP isotope ratio mass spectrometer (IRMS) at the Large Lakes Observatory (LLO). $\delta^{13}C$ of samples (DOC, POC, sediment OC and zooplankton) were determined at NOSAMS using an Optima IRMS on subsamples intended for radiocarbon analyses. Stable nitrogen isotope ratios ($\delta^{15}N$) and a set of samples for $\delta^{13}C$ of POM and $\delta^{13}C$ of zooplankton were measured at LLO using a Finnigan Delta Plus XP IRMS with Conflo III interface (Thermo Fisher Scientific Inc., Waltham, MA) coupled to a Costech ECS 4010 EA. Typical instrumental precisions of $\delta^{15}N$ and $\delta^{13}C$ based on analyses of multiple external standards were 0.17‰ and 0.2‰, respectively. The stable isotope ratios ($^{13}C$:$^{12}C$ and $^{15}N$:$^{14}N$) are reported as $\delta^{13}C$ and $\delta^{15}N$ respectively, which are the per mil difference relative to Vienna Pee Dee Belemnite carbonate and atmospheric air standards.

## 6.2.4 Bayesian MixSIR mixing model for multiple endmembers

The Bayesian isotopic modeling software MixSIR (Version 1.04) (Moore and Semmens 2008; Semmens et al. 2009) was used to partition the proportional contributions of potential OC sources to the bulk POC and to zooplankton diet based on their $\Delta^{14}C$ and $\delta^{13}C$ signatures. The MixSIR model works by determining probability distributions of sources contributing to the observed mixed signal while accounting explicitly for the uncertainty in the isotopic signatures of the sources and fractionation. The uncertainty of $\delta^{13}C$ and $\Delta^{14}C$ values used for modeling here are the analytical uncertainties based on analyses of multiple external standards or multiple analyses of graphite targets in the case of $\Delta^{14}C$. Since isotopic fractionation is already corrected for in the calculation of $\Delta^{14}C$ values, radiocarbon fractionation was not used in the model (thus specified as zero). Isotopic fractionation of +1 was used for $\delta^{13}C$ (Fry and Sherr, 1984). Prior information was not used in the model, hence all possible source combinations were equally possible contributions to the observed mixed signal. The number of iterations used was 10,000,000 (and 100,000,000 when the posterior draws were less than 1000). For each potential source, we report the median and the 5% and 95% confidence percentiles estimates of the proportional contribution of the sources to the measured (observed) value.

### 6.2.4.1. Choice of endmembers

To determine carbon sources to POC and food sources supporting mesozooplankton biomass, we chose isotopic end-members based on identifiable unique sources of OC to the POC pool in the lake (Zigah et al., 2011; Zigah et al., 2012). Because our modeling is based upon natural abundance stable carbon and radiocarbon distributions, these end-members vary from those generally used in labeling experiments (e.g., Taipale et al., 2008) or natural abundance stable isotope modeling (e.g., Karlsson et al., 2003). Based upon their unique $\Delta^{14}C$ values, the potential components of POC in the lake include carbon derived from recent photosynthesis (here described as "algal carbon" although it also includes herbivore biomass supported by recent primary production), bacterial OC, terrestrial OC, and resuspended-sediment OC. As this is a novel suite of end-members relative to previous work, we discuss our rationale further below.

Lake-wide primary production in Lake Superior is estimated at 9.73 Tg C per year (Sterner, 2010), although most of the POC input from phytoplankton photosynthesis is thought to be mineralized rapidly (Maier and Swain, 1978; Urban et al., 2005) and does not persist in the lake. The POC pool in the lake is only ~ 1 Tg C, (compared to ~15 Tg DOC and ~122 Tg DIC; Zigah et al., 2012). We assigned a $\delta^{13}C$ value of -30 ± 1‰ as representative of algal C (Sierszen et al. 2006). We used a $\Delta^{14}C$ of DIC as the $\Delta^{14}C$ of algal carbon from recent photosynthesis as DIC-incorporation is the starting point for algal biomass production (McNichol and Lihini, 2007; McCarthy et al., 2011). For calculating $\Delta^{14}C$ values, measured radiocarbon values are normalized to remove mass-dependent isotopic fractionation such that $\Delta^{14}C$ values reflect only time (age) or mixing (variably aged components). To normalize the sample, fractionation between $^{14}C$ and $^{12}C$ is assumed to be approximately twice that between $^{13}C$ and $^{12}C$ since the mass difference between $^{14}C$ and $^{12}C$ is twice that between $^{13}C$ and $^{12}C$ (Donahue et al., 1990; McNichol and Lihini, 2007). Therefore, in a modern system, the $\Delta^{14}C$ of algal carbon tracks that of DIC that was incorporated.

Bacterial carbon is another identifiable component of POC in the lake (Cotner et al., 2004). For $\Delta^{14}C$ and $\delta^{13}C$ of bacterial carbon, we used the $\Delta^{14}C$ and $\delta^{13}C$ of DOC, the main microbial food source, as we do not have direct measurement of bacterial biomass $\Delta^{14}C$ and $\delta^{13}C$. We acknowledge that this is only a first order approximation of the $\Delta^{14}C$ and $\delta^{13}C$ of bacteria in Lake Superior, and look forward to further refining this model endpoint when better data become available.

Radiocarbon values of bulk POC in the lake suggest that they contain a pre-aged carbon source that may result from sediment resuspension and further that this resuspension can impact surface water samples as well as deeper samples (Zigah et al., 2011; Zigah et al., 2012). This finding is consistent with previous studies showing the importance of sediment resuspension in Lake Superior (Urban et al., 2005; Churchill and Williams, 2004; Flood, 1989; Flood and Johnson, 1984). In our Lake Superior work (see below), the OC in the surface sediments (0-2 cm) at the various study stations across the lake have $\Delta^{14}C$ values that are older ($^{14}C$-depleted) than recent algal OC from lake photosynthesis. The physical mechanisms transporting such old OC from the lake sediments into the lake surface water are not well known. However, Lake Superior is dimictic, thus, there is density driven vertical mixing of the water column twice each year. Hence, organic materials resuspended into the deep waters due to strong bottom currents could be introduced into the surface waters during the lake overturn. In our lake surface (top 0-2 cm) sediment samples, the $\delta^{13}C$ values of OC were -27.3‰, -26.9‰, -27.0‰, -26.2‰, and -27.0‰, and the $\Delta^{14}C$ values were -20 ± 3‰, -37 ± 2‰, -23 ±

2‰, -117 ± 2‰, and -36 ± 2‰ at sites CM, SM, NM, WM, and EM respectively, and the corresponding site-specific values were used as the sediment OC end-member for each site in the lake. We note that this end member assignment is a first order approximation as lateral advection of old OC from shallower depths, especially at the WM site, is also possible.

The choice of end member for terrestrial OC was challenging because there are both 'old' and 'recent' terrestrial OC components. The radiocarbon value of recently synthesized terrestrial OC reflects the radiocarbon value of atmospheric $CO_2$. We determined the radiocarbon value of atmospheric $CO_2$ using the radiocarbon content of corn leaves (annual plant) collected in the western watershed of the lake ($\Delta^{14}C$ = 38 ± 2‰, Zigah et al., 2011). While this approach does not cover the entire watershed of the lake, we do not think there would be considerable differences across the basins because most variations in atmospheric $^{14}C$ occur at a larger spatial scale. The remoteness of the lake from big industrial plants or big cities, the uniformity of surface $\Delta^{14}C$-DIC across the lake, and the absence of considerable soot (black) carbon in the POC pools across the lake (Zigah et al., 2012) suggest little regional variation in atmospheric $^{14}C$ around Lake Superior. To account for the fact that terrestrially produced OC could spend some time in the soil before delivery to the lake, we performed additional model runs replacing the corn $\Delta^{14}C$ value with that of POC collected during high flow conditions in June 2008 from Amity Creek ($\Delta^{14}C$ = 11 ± 2‰; $\delta^{13}C$ = -27.3‰, Zigah et al., 2011), a primarily forested watershed north shore stream which drains into western Lake Superior. The choice of highflow data was because most terrestrial influx in streams and rivers occurs during storm flows. While we note that using data from one stream within the watershed might not be representative, the similarity of high flow Amity Creek POC $\Delta^{14}C$ to nearshore POC $\Delta^{14}C$ ($\Delta^{14}C$ range of 7-17‰) from both the southern and northern nearshore regions of the lake that we sampled implies that our terrestrial end-member POC-$\Delta^{14}C$ is a good first approximation.

### 6.2.5 Zooplankton allochthony based on $\Delta^{14}C$

Zooplankton allochthony in Lake Superior was estimated using a binary (terrigenous and autochthonous) mixing model as follows:

$$\Delta^{14}C_{\text{Zoop}} = f \bullet \Delta^{14}C_{\text{Terr}} + (1 - f) \bullet \Delta^{14}C_{\text{Algal}} \tag{1}$$

where f is the fraction of terrestrial OC in the zooplankton biomass, (1 – f) is the fraction of algal-derived carbon in the zooplankton biomass, and the subscripts 'Terr' and 'Algal' refer to terrestrial and algal-derived, respectively. We used $\Delta^{14}C$ of DIC as the algal-derived OC endmember. For the terrestrial endmember, we used the atmospheric $CO_2$ $\Delta^{14}C$ and $\Delta^{14}C$ of POC from high flow Amity Creek in separate model runs for sensitivity analysis.

### 6.2.6 Statistical analyses

We used SigmaPlot 9.0 (Systat Software Inc., San Jose, California, USA) for all statistical analyses. Relationships among samples were tested via correlation analyses in which case we report the Pearson's correlation coefficient (r), probability ($p$), and number of samples (n). The difference between isotopic composition of zooplankton and that of DIC, POC and DOC was determined using paired t-tests, and for these we reported the two-tailed probability value ($p$), and the number of samples (n). Significance difference or correlation was tested at 95% confidence level ($\alpha = 0.05$).

## 6.3. Results

### 6.3.1 Lake Superior isotopic distributions

The bulk POC in the lake (including both stratified and isothermal surface samples) had a mean $\Delta^{14}C$ value of 10 ± 29‰ (range -55‰ to 39‰, n = 14) (Fig. 2A and 2B; Table 2), and the $\Delta^{14}C$ of DOC in the lake was 38 ± 21‰ (range -10‰ to 74‰, n = 13) (Fig. 2A and 2B). $\Delta^{14}C$ of mesozooplankton varied from 36 to 38‰ at NB and ONT sites (both nearshore regions) to 62‰ at CM and SM (both offshore regions) (Fig. 2A and 2B). At each site $\Delta^{14}C$ of mesozooplankton and $\Delta^{14}C$ of DIC were similar (Fig. 2A and 2B) and a paired t-test showed no significant difference in their values ($p$ = 0.96, n = 13). In contrast, $\Delta^{14}C$ of mesozooplankton was significantly more positive (thus, $^{14}C$-enriched) than $\Delta^{14}C$ of either POC (t-test, $p < 0.0001$, n = 13) (Fig. 2A and 2B) or DOC (t-test, p = 0.03, n = 13) (Fig. 2A and 2B).

The $\delta^{13}C$ and $\delta^{15}N$ of POM in Lake Superior exhibited seasonal shifts. The bulk POM was more $^{13}C$-enriched ($\delta^{13}C$, mean = -28.2 ± 0.6‰, range -27.1‰ to -28.9‰, n = 7) and $^{15}N$-depleted ($\delta^{15}N$, mean = 0.5 ± 0.8‰, range -3.9‰ to -2.0‰, n = 7; Table 2) during stratification in August (excluding $\delta^{13}C$ of POM at EM) compared to the bulk POM in the isothermal lake in June ($\delta^{13}C$, mean = -29.9 ± 0.4‰, range -29.5‰ to -30.4‰, n = 7; $\delta^{15}N$, mean = -2.9 ± 0.6‰, range -0.6‰ to 1.7‰, n = 7; Table 2). In contrast, the stable isotopic composition of mesozooplankton in Lake Superior exhibited no seasonal shift for carbon and a smaller shift for nitrogen (Table 2). The $\delta^{13}C$ of mesozooplankton in Lake Superior was -30.0 ± 0.6‰ (range -29.5‰ to -31.2‰, n = 7) during isothermal conditions in June, and -30.0 ± 1.0‰ (range -28.2‰ to -31.3‰, n = 6) during stratification in August (excluding $\delta^{13}C_{POM}$ from ONT; Table 2). The average $\delta^{15}N$ of mesozooplankton in the isothermal lake in June was 5.0 ± 0.5‰ (range 4.2‰ to 5.4‰, n = 6), and shifted to 3.4 ± 0.8‰ (range 2.4‰ to 4.4‰, n = 6) in August (excluding ONT data in both seasons) (Table 2).

The $\delta^{15}N$ and $\delta^{13}C$ values of consumers reflect both the isotopic composition of the incorporated food plus biochemical fractionations. Movement across trophic levels imposes additional fractionation on the resulting biomass, with consumers exhibiting $^{13}C$-enriched values of ~0.5-1‰ (Fry and Sherr, 1984) and $^{15}N$-enriched values of 2-3‰ (Fry, 1991) relative to their food source. In Lake Superior, the mesozooplankton were $^{15}N$-enriched by an average of ~4‰ relative to bulk POM during isothermal condition, and $^{15}N$-enriched by ~6‰ relative to bulk POM during stratification (Table 2). In contrast to $\delta^{15}N$ values, mesozooplankton were $^{13}C$-depleted by an average of ~1‰ relative to bulk POM during stratification in August (Table 2).

The C:N values of mesozooplankton ranged from 6.2 to 8.7 (mean 7.1, n = 14), and were consistently and significantly lower than the C:N values (C:N, mean 8.2, range 7.0 to 9.5) of bulk POM (t test, $p = 0.001$, n = 14) (Table 2).

## 6.3.2. Modeling sources of POC and Mesozooplankton diet in Lake Superior

### 6.3.2.1. POC sources

The contribution of potential source materials to the bulk POC was estimated using the Bayesian MixSIR model based on source $\Delta^{14}C$ and $\delta^{13}C$ signatures. Based on the model results, the median (and 5% and 95% confidence percentiles) contribution of algal carbon to the bulk POM varied from 10% (5 – 14%) at the EM site during isothermal condition to

85% (77 – 93%) at the NM site during isothermal condition (Table 3). The median contribution of terrestrial carbon to bulk POM ranged from 1% (0.1 – 5%) at EM site during isothermal condition to 19% (2 – 47%) at the WM site during stratification (Table 3). Sedimentary OC influence on bulk POM varied from a median of 10% (2 – 15%) during stratification at WM site to 87% (84-91%) during isothermal condition at EM site (Table 3). The average lakewide (including both seasons) median contributions of algal, terrestrial, sedimentary, and bacterial OC to the bulk POM were 58%, 4%, 34%, and 2% (Table 3), and the corresponding values with creek POC as terrestrial endmember were 58%, 6%, 32%, and 3%, respectively (Table 3). It should be noted that these model estimates change to 23%, 28%, 15%, and 25%, respectively, if only $\Delta^{14}C$ values (rather than both $\Delta^{14}C$ and $\delta^{13}C$) are used in the MixSIR modeling.

### 6.3.2.2. Sources of mesozooplankton diet

After estimating the relative contributions of potential basal food resources to the bulk POM, we used the Bayesian MixSIR mixing model to evaluate the relative contributions of these basal foods to mesozooplankton production in Lake Superior. Algal-based food contributed a disproportionately large amount to the mesozooplankton biomass in Lake Superior, with a median contribution (average of both isothermal and stratified season at all sites) of 91% (range: 85 - 95%). If only $\Delta^{14}C$ values are used in the MixSIR modeling, the algal-based contribution changes to 40-89%. Based upon MixSIR modeling with both $\delta^{13}C$ and $\Delta^{14}C$ as inputs, mesozooplankton in Lake Superior also appear to gain a lakewide median of 3% (2- 4%) of their biomass carbon from consumption of bacterial biomass (Table 4). Although making up a considerable portion of the bulk POC in the lake, OC from the sediment and terrestrial POC contributed minimally (median: 3% each) to mesozooplankton carbon (Table 4).

Based on a two-endmember mixing model using $\Delta^{14}C$ values representing recent in-lake primary production and terrestrial inputs as the endmembers, mesozooplankton allochthony varied across the lake sites (Table 5), ranging from 0-54% (with corn leaves used as the terrestrial endmember) or 0-25% (with creek POC as the terrestrial endmember). The mesozooplankton autochthony estimates from both multiple endmember, multiple isotope (Table 4) and binary endmember radiocarbon-based (Table 5) models were comparable with a lakewide average offset of ~8% or 3% if the Isothermal SM site with large difference is excluded. The offset is only $\leq$ 1% when the creek POC is used as the terrestrial endmember. The mesozooplankton allochthony estimate from the binary model varied from the terrestrial contribution from the multiple endmember Bayesian model with a lakewide average offset of ~18% or 11% (if Isothermal SM is excluded). The offset, however, is ≤6% when creek POC is used as terrestrial endmember.

### 6.3.3 Cross-system comparisons of isotopic distributions (Hudson River, Small Lakes, Lake Superior, and the Pacific Ocean)

Zooplankton in the Hudson River had pre-aged radiocarbon content ($\Delta^{14}C$ = -236‰; Caraco et al., 2010), and were $^{14}C$-depleted relative to recent terrestrial OC, algae (based on $\Delta^{14}C_{DIC}$), POC, and DOC (Table 6). The $\Delta^{14}C$ of zooplankton in small lakes (Bran-de-Scie, Des Monts, Stukely, Bowker, and Fraser Lakes) ranged from -2‰ in Bowker Lake to 40‰ in Des Monts Lake (Fig. 3A), and was consistently $^{14}C$-depleted relative to concurrent POC, DOC, and DIC (with the exception of Fraser Lake with a DIC $\Delta^{14}C$ of -91‰ vs. a zooplankton $\Delta^{14}C$ of 10‰, Table 6; Fig. 3A). It is evident from $\Delta\Delta^{14}C_{DIC\text{-}Zoop}$ ($\Delta^{14}C_{DIC}$ - $\Delta^{14}C_{Zoop}$) vs. $\Delta\Delta^{14}C_{POC\text{-}Zoop}$ ($\Delta^{14}C_{POC}$ - $\Delta^{14}C_{Zoop}$) that the $\Delta^{14}C$ values of zooplankton in the small lakes are more similar to $\Delta^{14}C$ of DIC than $\Delta^{14}C$ of POC, with the exception of Fraser Lake where the absolute values of $\Delta\Delta^{14}C_{DIC\text{-}Zoop}$ and $\Delta\Delta^{14}C_{POC\text{-}Zoop}$ were almost the same (Table 6).

In the coastal Pacific Ocean, $\Delta^{14}C_{DIC}$ of 100‰ and $\Delta^{14}C_{POC}$ of 100‰ were identical and slightly $^{14}C$-enriched compared to $\Delta^{14}C_{zoop}$ of 76‰ (Fig. 3B; Table 6). A similar trend was observed in the north central Pacific Ocean where $\Delta^{14}C_{zoop}$ of 124‰ was slightly $^{14}C$-depleted relative to the $\Delta^{14}C_{DIC}$ of 132‰ and $\Delta^{14}C_{POC}$ of 139‰ (Fig. 3B; Table 6). In contrast, the zooplankton in the northeast Pacific Ocean had $\Delta^{14}C$ values identical to $\Delta^{14}C$ of POC, but different from the $\Delta^{14}C$ of DIC as evident in the $\Delta\Delta^{14}C_{DIC\text{-}Zoop}$ and $\Delta\Delta^{14}C_{POC\text{-}Zoop}$ of 69‰ and -4‰, respectively (Table 6).

The entire dataset was pooled to assess inter-system trends (thus small-to-large water body ecosystems) in $\Delta^{14}C_{zoops}$ vs. $\Delta^{14}C_{DIC}$, and $\Delta^{14}C_{zoops}$ vs. $\Delta^{14}C_{POC}$. There was a positive correlation between $\Delta^{14}C$ values of zooplankton and DIC (excluding Hudson River) in the pooled $\Delta^{14}C$ data ($r = 0.82$, $p < 0.0001$, $n = 14$) (Fig. 4A), with ~67% of the variation in $\Delta^{14}C$ of zooplankton accounted for by the changes in $\Delta^{14}C$ of phytoplankton utilizing in situ DIC (based on correlation coefficient of 0.67, Table 4A).

In contrast, $\Delta^{14}C$ of zooplankton was not correlated to $\Delta^{14}C$ of bulk POC (r = 0.03, $p$ = 0.92, n = 14) (Fig. 4B).

There was a positive correlation between $\Delta\Delta^{14}C_{DIC\text{-}Zoop}$ and the ratio of basin area to lake surface area (correlation, r = 0.88, $p$ = 0.047) (Fig. 5A). Although marginally significant (correlation, r = -0.84, $p$ = 0.078), there was a negative correlation between the hydrological residence time of the lakes and $\Delta\Delta^{14}C_{DIC\text{-}Zoop}$ (Fig. 5B).

## 6.4. Discussion

### 6.4.1 Composition of bulk POM, and putative food sources for consumers in Lake Superior

Isotopic signatures of baseline food resources can be used to assess their relative importance in the diet of their animal consumers. Food-source tracking using isotopic signatures works if a measurable contrast exists between the potential food resources. Mesozooplankton in Lake Superior could obtain their diet from recent primary production (perhaps cycled through an additional small herbivore first), bacterial biomass, terrestrial OC, or OC from resuspended sediments.

Our results indicate that in Lake Superior, the proportional median contribution of recent primary production to bulk POC was 58% (Table 3). This is not surprising for a large cold temperate oligotrophic lake with low levels of autochthonous primary production (Cotner et al., 2004; Sterner, 2010). Although the estimated annual lake-wide primary production is 9.73 Tg C (Sterner, 2010), the OC input from phytoplankton photosynthesis is thought to be mineralized rapidly (Maier and Swain, 1978; Urban et al., 2005) and does not persist in the lake. Consequently, the POC pool in the lake is estimated at only ~ 1 Tg C (compared to ~15 Tg DOC and ~122 Tg DIC; Urban et al., 2005; Zigah et al., 2012).

Our model estimates show that the combined proportions of terrestrial OC and resuspended-sediment OC can constitute a considerable fraction of the basal food available to consumers in the lake (Table 3). These estimates are consistent with published values from previous studies in the lake. Urban et al. (2004) reported that resuspended sedimentary OC contributed 10-35% of OC in sinking POC off the Keweenaw Peninsula at the depth of 25-35 m in the lake.

### 6.4.2 Stable isotopes and C:N ratios

Although bulk POM was $^{13}$C-enriched and $^{15}$N-depleted during stratification in August relative to isothermal conditions in June, the mesozooplankton in the lake did not exhibit such seasonal changes in their $\delta^{13}$C and $\delta^{15}$N signatures (Table 2). Mesozooplankton in the lake were generally $^{13}$C-depleted relative to the bulk POM, especially in the productive surface waters during stratification in August. Such $^{13}$C-depletion of mesozooplankton compared to bulk POM has been reported by several researchers (del Giorgio and France, 1996; Karlsson et al., 2003; Pulido-Villena et al., 2005; Mathews and Mazumder, 2006; McCallister and del Giorgio, 2008), and suggests that the mesozooplankton were primarily supported by a subsurface algal food with $^{13}$C-depleted values, or a baseline algal food source within the surface POM with $^{13}$C-depleted values as typical trophic-level enrichments for $\delta^{13}$C are +0.5 to +1‰ (Fry and Sherr, 1984). Another possibility is the accumulation and/or storage of lipids by the mesozooplankton from their food, thus making their entire biomass or whole body more $^{13}$C-depleted than their food source as lipids are more $^{13}$C-depleted than other biochemicals in their biomass (DeNiro and Epstein, 1978; McConnaughey and McRoy, 1979; Kling et al., 1992; Matthew and Mazumder, 2005; Smyntek et al., 2007). Mesozooplankton in Lake Superior do exhibit an increase in C:N values during stratification in August relative to isothermal conditions in June, which is consistent with increasing accumulation and storage of lipids during the more productive and warmer season (McConnaughey and McRoy, 1979; Kiljunen et al., 2006; Smyntek et al., 2007; Syvaranta and Rautio, 2010).

### 6.4.3 Radiocarbon-based estimation of mesozooplankton food sources in Lake Superior

Taken together, and without seasonal comparison, the stable C and N isotope values do not distinguish mesozooplankton from the bulk POM pool from which they feed, especially during isothermal conditions in the lake. Adding radiocarbon information allows for the estimation of mesozooplankton dependence on food sources other than that year's in situ primary production (and its immediate consumers), and also significantly refines the relationship between POM and mesozooplankton.

Mesozooplankton in Lake Superior in both isothermal and stratified conditions have $\Delta^{14}$C values that track those of co-occurring DIC rather than bulk POM (Fig. 1), indicating that the mesozooplankton in this system are preferentially feeding on food resources resulting from contemporary photosynthesis rather than indiscriminately upon bulk POM. Bayesian MixSIR modeling results generally show that most of the mesozooplankton biomass in the entire lake, and in both seasons (medians 85-95%; Table 4) came from incorporation of recent primary production. These results are generally consistent with mesozooplankton autochthony estimates from binary isotopic mixing modeling with the exception of SM site during isothermal conditions (range 61-100% or 75-100% depending on choice of terrestrial endmember as shown in Table 5). Both approaches show considerable enrichment in mesozooplankton biomass relative to the proportion of "algae" in bulk POC (median, 58%; Table 3). That algal carbon dominantly supports mesozooplankton biomass production was not surprising as algal-derived food is generally known to be labile and the most preferred food option for secondary producers (Brett et al., 2009). Our results agree with previous studies in other lakes (del Giorgio and France, 1996; Cole et al., 2002; McCallister and del Giorgio, 2008; Mohammed and Taylor, 2009; Karlsson et al., 2012) and rivers (Sobczak et al., 2002; Thorp and Delong, 2002; Meersche et al., 2009) that reported that zooplankton were sustained disproportionately and/or largely by phytoplankton biomass.

Mesozooplankton dependence on organic carbon subsidies (sedimentary and terrestrial OC) in Lake Superior was small (Table 4), although these organic carbon resources make up a considerable fraction of the bulk POC in the lake (Table 3). Contrary to our results, other studies have reported larger use of non-algal food by zooplankton in some aquatic systems based on either natural abundances of $\Delta^{14}C$ (Schell, 1983; Caraco et al., 2010), $\delta^{13}C$ and $\delta^{15}N$ (Meili et al., 1996; Jones et al., 1998; Karlsson et al., 2004; Matthews and Mazumber, 2006), or whole lake addition of $^{13}C$-labeled bicarbonates (Carpenter et al., 2005; Pace et al., 2007; Taipale et al., 2008).

Differently aged components (modern vs. ancient) of organic carbon subsidies may have different fates in aquatic ecosystems. The relative ages of the non-algal OC that support heterotrophic microbial communities and the upper trophic levels of food webs are not well known although this knowledge is essential in understanding food web dynamics. In Lake Superior, although pre-aged organic carbon from the sediment was a putative food option in the lake, and constituted a median proportion of as much as 87% (84-91%) of the available food carbon (POC) during isothermal (mixed-lake water) condition at EM site and 84% (78-87%)

during stratified condition at SM site (Table 3), mesozooplankton in the lake only incorporated trace amounts (median: 3% [1-7%]) of this old carbon into their biomass (Table 4). This observation could be due to a general decrease in palatability of considerably aged organic carbon or could be due to the extensive amount of reworking this material has experienced in Lake Superior. Some studies have suggested that modern terrestrial organic carbon supports heterotrophic respiration (Mayorga et al., 2005) whereas ancient terrestrial components could be important food sources for heterotrophic microbes (McCallister et al., 2004) and animal consumers (Ishikawa et al., 2010) in certain aquatic systems. In contrast to Lake Superior, studies of the Hudson River food-web (Caraco et al., 2010) and bacterial biomass production in the Hudson and York River systems (McCallister et al., 2004) have shown that both mesozooplankton and bacteria can use considerably aged reduced carbon as a food source. Also, in the open ocean in eastern North Pacific, radiocarbon studies show that bacteria assimilate both modern and ancient organic carbon (Cherrier et al., 1999). Schell (1983) in a study of the Colville River and coastal Alaskan Beaufort Sea reported that old carbon from peat in the catchment introduced primarily into food webs in the freshwater portions of the system, i.e., anadromous fish and ducks feeding in these areas. While it is still not clear which aquatic variables drive the relative utilization of ancient vs. modern food sources in these systems, some studies have indicated that terrestrial materials from the catchment are less refractory than previously thought (Hessen, 1992; Tranvik, 1992), and others have suggested addition of new synthesized algal food could act as co-metabolic primer facilitating the use of the aged (potentially refractory) organic material (Horvath, 1972; McCallister et al., 2004; Goni et al., 2006; Aller et al., 2008).

It should be noted that, while this study is one of the most extensive isotopic (particularly radiocarbon) investigations of the ecosystem of any of the great lakes in the world, our results represent a general large-scale view of the ecosystem functioning of Lake Superior since the spatial and temporal coverage of this study is limited to 8 sampling sites covering nearshore and offshore locations, and visited twice during thermal stratification and mixed-lake condition. A high resolution spatial and temporal sampling scheme would be needed for a more detailed understanding of the feeding habits and/ ecology of the mesozooplankton in the lake.

### 6.4.4 Comparison of zooplankton food sources in small-to-large aquatic systems

To gain cross-system insight into the food sources supporting animal consumers in aquatic systems, we compared the food sources of zooplankton in the Hudson River, five separate small northern lakes, and different sites in the North Pacific Ocean, to the food resources supporting zooplankton in a large lake (Lake Superior). This cross-system dataset is not representative of global lake diversity and/or variability and are only from North America as we cannot find radiocarbon composition of zooplankton in aquatic systems in other parts of the world. Trends observed and discussed here give a broad picture of ecosystem functioning across lake size gradient in the US and Canada. A more globally distributed dataset is needed to ascertain whether the trends observed in this study are consistent with the global view of the relationship between lake-size and zooplankton ecology.

In the Hudson River, and Bran-de-Scie, Des Monts, Stukely, and Bowker Lakes, the zooplankton biomass was generally largely supported by in-situ primary production (and its immediate consumers) as evidenced by smaller values of $\Delta\Delta^{14}C_{DIC\text{-}Zoop}$ relative to $\Delta\Delta^{14}C_{POC\text{-}Zoop}$ (Table 6). However, the $^{14}C$-depletion of zooplankton biomass relative to the putative autochthonous food sources (Table 6) indicates the use of some aged allochthonous food source by the zooplankton for their dietary needs. Zooplankton incorporation of aged allochthonous food in these small lake systems contrasts with observations in Lake Superior, where the mesozooplankton preferentially and heavily depended on in situ primary production. Conservative estimates based on $\Delta\Delta^{14}C_{DIC\text{-}Zoop}$ indicate that the proportion of allochthonous food supporting zooplankton in the small lakes (except Fraser Lake) was larger than that in open Lake Superior (Table 6). This is also consistent with the observed relationship between zooplankton dependence on allochthonous food resources and variables such as ratio of catchment area to lake surface area (Fig. 5A), and water residence time (Fig. 5B). The ratio of basin area to surface area of a lake gives an indication of potential terrestrial subsidy to the lake's ecosystem. As the basin area-to-surface area ratio increases, suggesting potentially higher terrestrial influence, the difference between $\Delta^{14}C_{Zoop}$ and $\Delta^{14}C_{DIC}$ also increases as reflected in the positive correlation between $\Delta\Delta^{14}C_{DIC\text{-}Zoop}$ and the ratio of basin area to lake surface area (Fig. 5A). Hydrologic residence time is a variable that is related to lake size. Small lakes tend to have shorter water residence times whereas large lakes usually hold water for longer time periods (Table 1). There was generally a negative relationship between the hydrological residence time of the lakes and $\Delta\Delta^{14}C_{DIC\text{-}Zoop}$ (Fig. 5B), implying the difference between $\Delta^{14}C$ of zooplankton and $\Delta^{14}C$ of DIC decreases with an increase in lake water residence time, and by extension, with lake size.

In the oceanic sites, $\Delta^{14}C$ values of zooplankton and bulk POM were similar at all sites (Fig. 3B), suggesting that either the bulk POM was almost entirely derived from that year's primary production, or that the zooplankton were indiscriminately feeding on the bulk POM. It is worth noting however that estimating zooplankton food sources in the oceanic sites is complicated by the considerable differences in $\Delta^{14}C$ values of DIC with depth and laterally, such that water mass movements, and migratory feeding of zooplankton (and upper trophic organisms) could significantly mask the actual radiocarbon relationships between zooplankton, DIC and POC. The pooled data from the small lakes, Lake Superior and the Pacific Ocean show strong correlation between $\Delta^{14}C$ values of zooplankton and DIC, but poor correlation between $\Delta^{14}C$ of zooplankton and bulk POM (Fig. 4A, 4B) indicating that in most aquatic ecosystems, recent in-situ primary production is the most preferred food resource for zooplankton.

It is worth stating that different zooplankton groups do have different feeding and/or ecological strategies, and the observed zooplankton food preferences and the relationship between lake size and allochthony of zooplankton discussed above could be influenced by this. While the higher mesozooplankton autochthony seen in Lake superior relative to the smaller lakes could be attributed to the specific filter feeding style of the predominant calanoid copepods (~70% of crustacean zooplankton) in the lake, the small offset between $\Delta^{14}C$ of algae (based on $\Delta^{14}C$-DIC) and $\Delta^{14}C$ of the bulk mesozooplankton suggest the remaining zooplankton groups in Lake Superior (~30%) including cyclopoid copepods and daphnids were also largely feeding on algae, although these zooplankton groups are adapted to utilize other food options such as detritus, protists, bacteria and other zooplankton. Since cladocerans such as daphnids, and cyclopoid copepods are typically more abundant in small nutrient enriched aquatic systems (Gannon and Stember, 1978; Balcer et al., 1984), it could be argued that their adaptation for feeding on non-algal food options is responsible for the relatively high zooplankton allochthony seen in the smaller lakes. However, the zooplankton composition itself could be coupled to nutrient (N and P) availability (Demott and Gulati, 1999; Schulz and Sterner, 1999; Conde-Porcuna et al., 2002) and hence terrestrial influence, consistent with the observed pattern of increasing zooplankton allochthony with terrestrial influence as seen in this study.

## 6.5. Conclusion

Our isotopic investigation shows that intermediate trophic-level mesozooplankton in Lake Superior prefer to incorporate fresh autochthonous food, despite the availability of other organic carbon sources, and that upper trophic levels are likely not supported by terrestrial and/or resuspended-sediment OC subsidies to the carbon cycle. A similar trend is apparent in our oceanic sites, and selected small lakes, although the small lakes do exhibit a higher degree of zooplankton dependence on allochthonous food resources. This provides real-world support to lab studies showing preferential incorporation of phytoplankton fatty acids and POC into herbivorous zooplankton offered mixed diets of terrestrially-derived and phytoplankton-derived particulate organic matter (Brett et al., 2009) and confirms in a large lake, and similar clear-water systems such as the open oceans, the observation from unproductive small lakes that zooplankton selectively incorporate fresh autochthonous organic carbon (Karlsson, 2007; Karlsson et al., 2012). Our results suggest that if spatial or temporal subsidies of organic carbon fuel the net heterotrophy seen in Lake Superior, their effects are limited to the microbial loop and lower trophic levels, and do not extend to mesozooplankton and higher trophic levels. Further research should focus upon catabolic metabolism of mesozooplankton and both anabolic and catabolic metabolism in the microbial loop to further our understanding of such subsidies in the carbon cycle and energy transfer.

## Acknowledgements

The authors thank Koushik Dutta, Meg MacDonald, and Danielle Schminkey from University of Minnesota Duluth, Annie Stuart and Lindsey Koren from Virginia Commonwealth University and the captain and crew of the R/V *Blue Heron* for sampling assistance; the staff at the National Ocean Sciences Accelerator Mass Spectrometry Facility (NOSAMS) for radiocarbon analyses; Koushik Dutta of the Organic and Stable Isotope Biogeochemistry Analytical Research Facility (OSI-BARF) at the Large Lakes Observatory, University of Minnesota Duluth for stable carbon and nitrogen isotope analyses; Robert Hecky of Large Lakes Observatory, University of Minnesota Duluth for pre-cruise discussions on zooplankton sampling, and comments on an earlier version of the manuscript, the associate editor Steven Bouillon and two anonymous reviewers whose valuable comments considerably improved this paper.

This work was funded by a National Science Foundation-Ocean Sciences grant (OCE- 0825600) to E.C.M. and J.P.W, and OCE-0825403 to S.L.M.

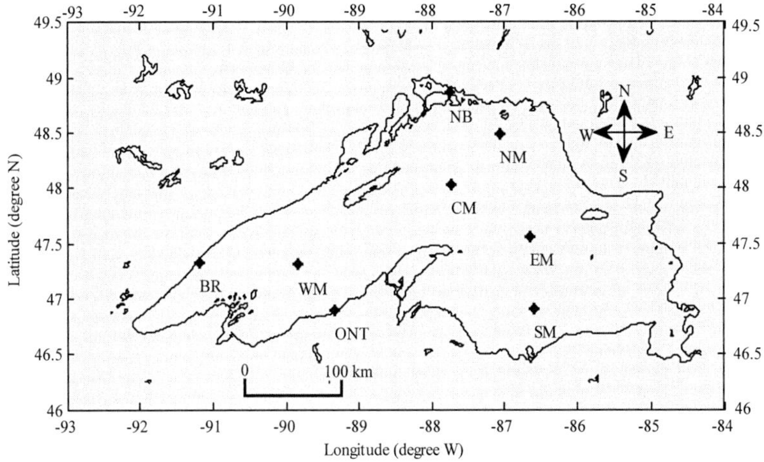


Figure 1. Map of Lake Superior showing sampling sites across the lake. The open lake sites include western station/mooring (WM), central mooring (CM), northern mooring (NM), eastern mooring (EM), and southern mooring (SM). The nearshore sites are off Baptism River (BR), off Ontonagon River (ONT), and off Nipigon Bay (NB).

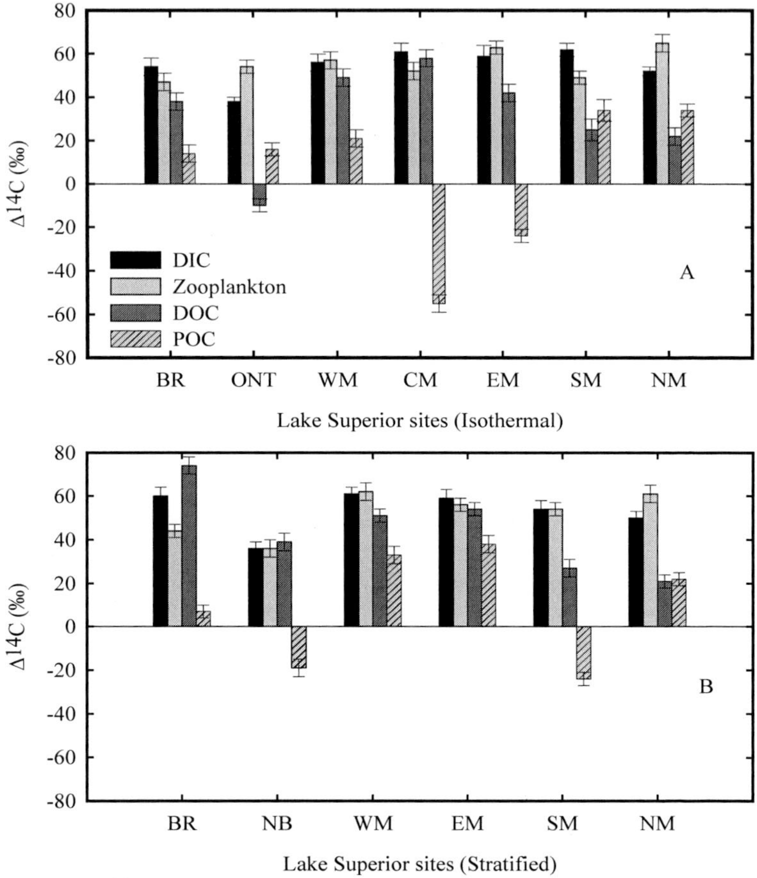


Figure 2. $\Delta^{14}C$ values for DIC, Mesozooplankton, DOC, and POC samples in (A) the isothermal lake in June, and (B) the stratified lake in August 2009.

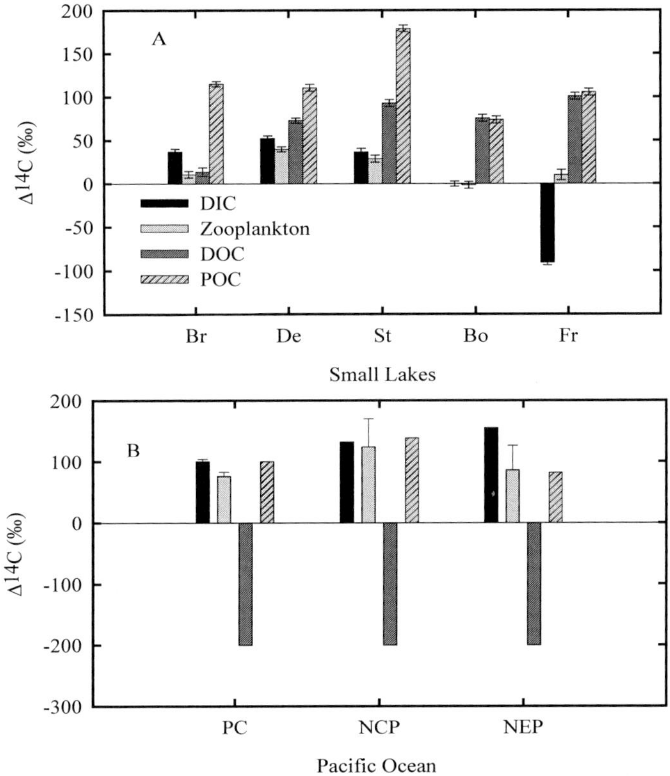


Figure 3. $\Delta^{14}C$ values for DIC, Zooplankton, DOC, and POC samples in (A) selected small lakes that includes Bran-de-Scie (Br), Des Monts (De), Stukely (St), Bowker (Bo), and Fraser (Fr) Lakes, and (B) sites in the Pacific Ocean that includes Pacific coastal ocean (PC), North Central Pacific (NCP), and North Eastern Pacific (NEP). Data adapted from William et al., 1987, Druffel and William, 1990, Druffel and William, 1991, Druffel et al., 1996.

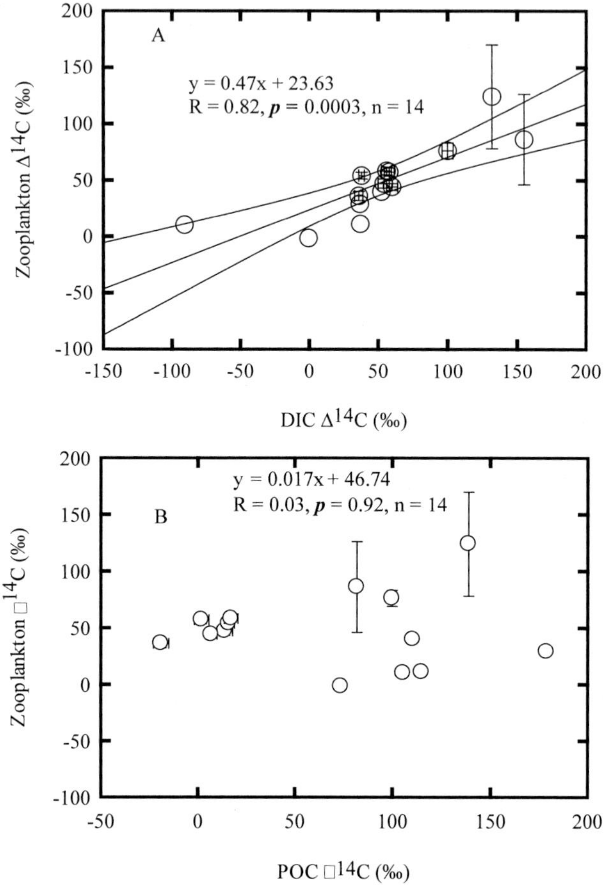


Figure 4. The relationship between $\Delta^{14}C$ of zooplankton and (A) $\Delta^{14}C$ of DIC, and (B) $\Delta^{14}C$ of POC. These comparisons show that the radiocarbon signatures of zooplankton are largely determined by those of DIC, and hence algal carbon from recent photosynthesis within the lake. For Lake Superior, two points, the average isothermal values and average stratification values, were used in order to not bias the trends.

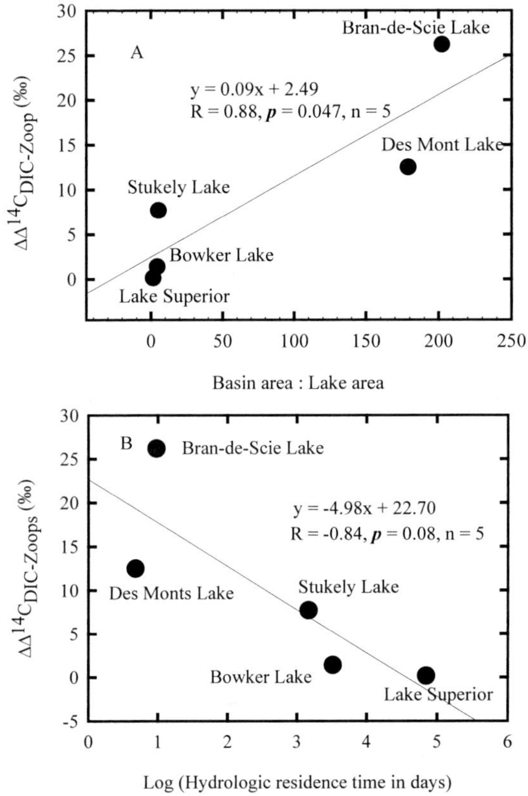


Figure 5. Relationship between the difference of $\Delta^{14}C$ values of DIC and Zooplankton, and (A) ratio of basin area to lake surface area, and (B) hydrologic residence time of the various lakes. Although marginally significant correlation (***p*** = 0.08) is observed between $\Delta\Delta^{14}C_{\text{DIC-Zoop}}$ and hydrologic residence time, these two plots generally illustrate that zooplankton support by allochthonous organic carbon is related to variables that indicate terrestrial influence.

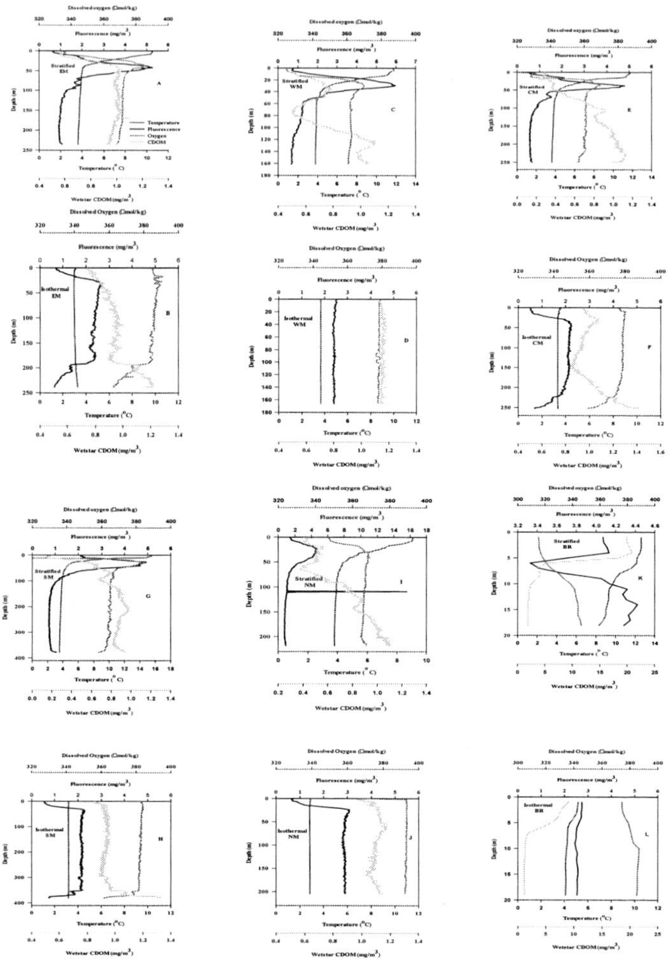

Supplementary Figure 1: CTD data of sampling sites in Lake Superior showing the depth profiles of temperature, beam transmission (Wetlab CStar,%), fluorescence (Wetlab Wetstar, mg/m$^3$), dissolved oxygen, and wet CDOM (Wetlab CDOM, mg/m$^3$) in the mixed water column in June and stratified condition in August.

**Table 6-1. Coordinates, sampling depths, hydrologic residence, basin area, and surface areas of the rivers, small lakes, Lake Superior and the Pacific Ocean. nd = not determined or not known.**

| Aquatic system | Water column condition | Date of samplin g | Latitude | Longitude | Water depth (m) | Sampling depth (m) | Hydrologic residence Time (years) | Surface Area ($km^2$) | Basin area ($km^2$) | Basin area / Surface area |
|---|---|---|---|---|---|---|---|---|---|---|
| Hudson river | Isothermal (mixed) | 2004-2005 | 40°.42 N-44°.06 N | 73 °56 W-74 °01 W | 7 | 0.2 | 0.3 | 760.41 | 34628.1 | 45.54* |
| Bran-de-Scie | Stratified (thermally) | 1 Sep 2004 | 45°.41 N | 72 °20 W | 8.4 | 0.5-1 | 0.026 | 0.13 | 26.3 | 202.31[&] |
| Des Monts | Stratified | 8 Sep 2004 | 45° 40 N | 72 °18 W | 5.5 | 0.5-1 | 0.013 | 0.26 | 46.5 | 178.85[&] |
| Stukely | Stratified | 15 Sep 2004 | 45°38 N | 72° 25 W | 33.1 | 0.5-1 | 4.03 | 4 | 20.8 | 5.20[&] |
| Bowker | Stratified | 15 Sep 2004 | 45 °41 N | 72° 22 W | 60.4 | 0.5-1 | 8.96 | 2.5 | 10.9 | 4.36[&] |
| Fraser | Stratified | 28 Sep 2004 | 45° 39 N | 72 °18 W | 18.7 | 0.5-1 | 0.36 | 1.6 | 61.8 | 38.63[&] |
| Baptism river mouth (BR) | Isothermal | 21 Jun 2009 | 47°33 N | 91°19 W | 20 | 0-15 | 191 | 82170 | 127700 | 1.55 |
| Baptism river mouth (BR) | Stratified | 24 Aug 2009 | 47°33 N | 91°19 W | 20 | 0-15 | 191 | 82170 | 127700 | 1.55 |
| Ontonagon river mouth (ONT) | Isothermal | 19 Jun 2009 | 46°9 N | 89°34 W | 20 | 0-10 | 191 | 82170 | 127700 | 1.55 |
| Nipigon Bay (NB) | Stratified | 16 Aug 2009 | 48°86 N | 87°76 W | 62 | 0-50 | 191 | 82170 | 127700 | 1.55 |
| Western Lake Superior (WM) | Isothermal | 20 Jun 2009 | 47°31 N | 89°85 W | 171 | 0-50 | 191 | 82170 | 127700 | 1.55 |
| Western Lake Superior (WM) | Stratified | 23 Aug 2009 | 47°31 N | 89°85 W | 171 | 0-50 | 191 | 82170 | 127700 | 1.55 |
| Central Lake Superior (CM) | Isothermal | 15 Jun 2009 | 48°03 N | 87°74 W | 257 | 0-50 | 191 | 82170 | 127700 | 1.55 |

| Eastern Lake Superior (EM) | Isothermal | 17 Jun 2009 | 47°56 N | 86°65 W | 242 | 0-50 | 191 | 82170 | 127700 | 1.55 |
|---|---|---|---|---|---|---|---|---|---|---|
| Eastern Lake Superior (EM) | Stratified | 15 Aug 2009 | 47°56 N | 86°65 W | 242 | 0-50 | 191 | 82170 | 127700 | 1.55 |
| Southern Lake Superior (SM) | Isothermal | 14 Jun 2009 | 46°91 N | 86°6 W | 398 | 0-50 | 191 | 82170 | 127700 | 1.55 |
| Southern Lake Superior (SM) | Stratified | 19 Aug 2009 | 46°91 N | 86°6 W | 398 | 0-50 | 191 | 82170 | 127700 | 1.55 |
| Northern Lake Superior (NM) | Isothermal | 16 Jun 2009 | 48°49 N | 87°06 W | 216 | 0-50 | 191 | 82170 | 127700 | 1.55 |
| Northern Lake Superior (NM) | Stratified | 16 Aug 2009 | 48°49 N | 87°06 W | 216 | 0-50 | 191 | 82170 | 127700 | 1.55 |
| Pacific coastal ocean | Stratified | 1980,19 86, 1987 | 27°33.0 N; 33°50 N | 114°52.3 W; 118°50. W | ~900 | 0-20 | 37000 | 1652000 00 | 20300000 | 0.12** |
| North Central Pacific (NCP) | Stratified | 1972-1983 | ~30°39'N; 31°00'N | 155°23W; 159°00W | ~5800 | 0-1700 | 37000 | 1652000 00 | 20300000 | 0.12[#] |
| Northeast Pacific (NEP) | Stratified | 1975-1977 | 32°34 N; 34°00 N | ~120°45 W; 123°00 W | ~4100 | 0-1500 | 37000 | 1652000 00 | 20300000 | 0.12[+] |

* Caraco et al. 2010, [&] McCallister and del Giorgio 2008, ** Druffel and William 1990 and Druffel and William 1991, [#] William et al. 1987 and Druffel and William 1990

[+] William et al. 1987 and Druffel et al. 1996

**Table 6-2. Radiocarbon, stable isotope and atomic C:N ratio of zooplankton (Zoop) and organic and inorganic carbon pools in the surface waters of Lake Superior.**

| | $\Delta^{14}C$ (‰) | | | | $\delta^{13}C$ (‰) | | | $\delta^{15}N$ (‰) | | Molar C/N | |
|---|---|---|---|---|---|---|---|---|---|---|---|
| Station and date | DIC | Zoop | DOC | POC | Zoop | DOC | POC | POC | Zoop | POC | Zoop |
| Jun-09 Isothermal | | | | | | | | | | | |
| WM | 56 ± 4 | 57 ± 4 | 49 ± 4 | 21 ± 4 | -29.5 ± 0.3 | -26.5 | -29.5 ± 0.4 | 1.7 | 5.4 | 7.2 | 6.4 |
| CM | 61 ± 4 | 52 ± 4 | 58 ± 4 | -55 ± 4 | -29.9 ± 1.1 | -25.9 | -29.9 ± 0.1 | 0.1 | 4.6 | 7.9 | 6.2 |
| EM | 59 ± 5 | 63 ± 3 | 42 ± 4 | -24 ± 3 | -30.0 ± 1.0 | -26.3 | -30.2 ± 0.2 | 0.1 | 4.9 | 7.9 | 6.7 |
| SM | 62 ± 3 | 49 ± 3 | 25 ± 5 | 34 ± 5 | -29.7 ± 0.0 | -26.0 | -29.8 ± 0.0 | 0.4 | 4.2 | 8 | 7.1 |
| NM | 52 ± 2 | 65 ± 4 | 22 ± 4 | 34 ± 3 | -30.0 ± 1.1 | -26.5 | -30.4 ± 0.7 | 0.7 | 5.4 | 7 | 6.2 |
| ONT | 38 ± 2 | 54 ± 3 | -10 ± 3 | 16 ± 3 | -31.2 ± 0.0 | nd | -29.5 ± 0.3 | -0.6 | 2.8 | 9.1 | 7.7 |
| BR | 54 ± 4 | 47 ± 4 | 38 ± 4 | 14 ± 4 | -29.5 ± 1.8 | nd | -30.4 ± 0.4 | 1.1 | 5.4 | 9.5 | 6.5 |
| Aug-09 Stratified | | | | | | | | | | | |
| WM | 61 ± 3 | 62 ± 4 | 51 ± 3 | 33 ± 4 | -30.3 ± 0.9 | -26.1 | -27.8 ± 0.0 | -2.8 | 3.6 | 8.4 | 7.6 |
| CM | 62 ± 3 | nd | nd | 39 ± 3 | nd | -26.0 | -28.8 ± 0.5 | -2.3 | Nd | 8.3 | nd |
| EM | 59 ± 4 | 56 ± 3 | 54 ± 3 | 38 ± 4 | -30.1 ± 0.5 | -26.0 | -30.2 ± 1.4 | -3.9 | 3.9 | 7.6 | 7.3 |
| SM | 54 ± 4 | 54 ± 3 | 27 ± 4 | -24 ± 3 | -29.4 ± 0.5 | -26.0 | -28.1 ± 0.0 | -2.9 | 3.4 | 8.4 | 6.5 |
| NM | 50 ± 3 | 61 ± 4 | 21 ± 3 | 22 ± 3 | -30.4 ± 0.5 | -26.1 | -28.9 ± 0.0 | -2.9 | 4.4 | 8.5 | 8.2 |
| ONT | 56 ± 4 | nd | nd | nd | -25.9 ± 0.4 | -28.3 | -27.2 ± 0.0 | -3.7 | 0.1 | 8.3 | 8.1 |
| BR | 60 ± 4 | 44 ± 3 | 74 ± 4 | 7 ± 3 | -28.2 ± 0.2 | -26.5 | -28.4 ± 0.1 | -2.7 | 2.6 | 8.4 | 5.8 |
| NB | 36 ± 3 | 36 ± 4 | 39 ± 4 | -19 ± 4 | -31.3 ± 0.8 | -26.5 | -28.1 ± 0.2 | -2 | 2.4 | 7.9 | 8.7 |

**Table 6-3. Relative contributions of various food resources to the bulk POM in the water column of Lake Superior. IsoSource model was used for these calculations with algal carbon from recent photosynthesis, terrestrial POC, sedimentary carbon, and bacterial biomass as end-members.**

| Station | Condition | Algal Carbon (%) | | | Terrestrial POC (%) | | | Sedimentary carbon (%) | | | Bacterial biomass (%) | | |
|---|---|---|---|---|---|---|---|---|---|---|---|---|---|
| | | Mean | Max | Min | Mean | Max | Min | Mean | Max | Min | Mean | Max | Min |
| CM | Stratified | 34.4 ± 18.7 | 73 | 0 | 18.2 ± 11.5 | 47 | 0 | 11.5 ± 7.0 | 29 | 0 | 35.9 ± 22.8 | 85 | 0 |
| EM | Isothermal | 4.1 ± 3.3 | 13 | 0 | 8.8 ± 6.6 | 27 | 0 | 82.0 ± 3.2 | 88 | 73 | 5.2 ± 4.0 | 16 | 0 |
| EM | Stratified | 33.5 ± 20.2 | 70 | 0 | 20.1 ± 13.6 | 45 | 0 | 12.3 ± 8.2 | 25 | 0 | 34.2 ± 21.5 | 75 | 0 |
| NM | Isothermal | 57.2 ± 7.7 | 76 | 40 | 14.9 ± 10.8 | 44 | 0 | 7.8 ± 5.9 | 24 | 0 | 20.1 ± 14.7 | 60 | 0 |
| NM | Stratified | 30.8 ± 12.3 | 63 | 0 | 24.0 ± 17.4 | 74 | 0 | 12.6 ± 9.3 | 39 | 0 | 32.5 ± 23.3 | 100 | 0 |
| SM | Isothermal | 46.7 ± 10.1 | 72 | 22 | 18.4 ± 13.4 | 56 | 0 | 9.3 ± 6.9 | 29 | 0 | 25.6 ± 18.5 | 78 | 0 |
| SM | Stratified | 4.5 ± 3.5 | 14 | 0 | 9.0 ± 6.7 | 28 | 0 | 79.8 ± 2.9 | 87 | 72 | 6.7 ± 5.1 | 21 | 0 |
| WM | Isothermal | 28.2 ± 18.9 | 79 | 0 | 31.4 ± 21.5 | 78 | 0 | 10.9 ± 5.3 | 20 | 0 | 10.9 ± 20.0 | 81 | 0 |
| WM | Stratified | 34.2 ± 20.4 | 84 | 0 | 23.6 ± 15.3 | 58 | 0 | 7.1 ± 4.1 | 16 | 0 | 35.0 ± 22.2 | 89 | 0 |

**Table 6-4. Contribution of putative food sources to zooplankton biomass in Lake Superior. The IsoSource model was used in computing these values using algal carbon from recent photosynthesis, terrestrial POC, sedimentary carbon, and bacterial biomass are food options.**

| Station | Condition | Algal Carbon (%) | | | Terrestrial POC (%) | | | Sedimentary carbon (%) | | | Bacterial biomass (%) | | |
|---|---|---|---|---|---|---|---|---|---|---|---|---|---|
| | | Mean | Max | Min | Mean | Max | Min | Mean | Max | Min | Mean | Max | Min |
| CM | Isothermal | 61.4 ± 17.4 | 90 | 10 | 6 ± 4.6 | 20 | 0 | 3.5 ± 2.8 | 12 | 0 | 29.1 ± 21 | 90 | 0 |
| EM | Isothermal | 89.2 ± 4.0 | 97 | 78 | 2.7 ± 2.3 | 9 | 0 | 1.2 ± 1.2 | 5 | 0 | 6.9 ± 5.4 | 22 | 0 |
| EM | Stratified | 71.7 ± 17.5 | 97 | 20 | 2.2 ± 1.9 | 8 | 0 | 0.9 ± 1.0 | 4 | 0 | 25.2 ± 18.7 | 80 | 0 |
| SM | Isothermal | 75.1 ± 5.0 | 87 | 63 | 8.6 ± 6.5 | 27 | 0 | 4.2 ± 3.3 | 14 | 0 | 12.1 ± 8.9 | 37 | 0 |
| SM | Stratified | 98.3 ± 1.0 | 100 | 97 | 1.0 ± 0.6 | 2 | 0 | 0.1 ± 0.3 | 1 | 0 | 1.0 ± 1.0 | 3 | 0 |
| WM | Isothermal | 100 ± 0 | 100 | 100 | 0 | 0 | 0 | 0 | 0 | 0 | 0 | 0 | 0 |
| WM | Stratified | 100 ± 0 | 100 | 100 | 0 | 0 | 0 | 0 | 0 | 0 | 0 | 0 | 0 |

**Table 6-5. Comparison of the radiocarbon signatures of DIC, DOC, POC, and zooplankton in small-to-large surface area aquatic systems.**

| Aquatic system | $\Delta^{14}C$-DIC | $\Delta^{14}C$-DOC | $\Delta^{14}C$-POC | $\Delta^{14}C$-Zoop | $\Delta\Delta^{14}C$ (DIC-Zoop) | $\Delta\Delta^{14}C$ (POC-Zoop) | Reference |
|---|---|---|---|---|---|---|---|
| | (‰) | (‰) | (‰) | (‰) | (‰) | (‰) | |
| Hudson river | -57 ± 14 | 40 ± 9 | -145 ± 27 | -236 ± 25 | 179 ± 29 | 91 ± 37 | Caraco et al. 2010 |
| Bran-de-Scie | 37 ± 3 | 14 ± 5 | 115 ± 3 | 11 ± 4 | 26 ± 5 | 104 ± 5 | McCallister and del Giorgio 2008 |
| Des Monts | 52 ± 3 | 73 ± 3 | 111 ± 4 | 40 ± 3 | 13 ± 4 | 71 ± 5 | McCallister and del Giorgio 2008 |
| Stukely | 37 ± 4 | 93 ± 4 | 179 ± 4 | 29 ± 4 | 8 ± 6 | 150 ± 6 | McCallister and del Giorgio 2008 |
| Bowker | -0.4 ± 3 | 76 ± 4 | 74 ± 4 | -2 ± 4 | 1 ± 5 | 76 ± 6 | McCallister and del Giorgio 2008 |
| Fraser | -90.9 ± 3 | 101 ± 4 | 106 ± 4 | 10 ± 6 | -101 ± 7 | 96 ± 7 | McCallister and del Giorgio 2008 |
| Baptism river mouth (BR) | 54 ± 4 | 38 ± 4 | 14 ± 4 | 47 ± 4 | 7 ± 6 | -33 ± 6 | This study |
| Baptism river mouth (BR) | 60 ± 4 | 74 ± 4 | 7 ± 3 | 44 ± 3 | 16 ± 5 | -37 ± 4 | This study |
| Ontonagon river mouth (ONT) | 38 ± 2 | -10 ± 3 | 16 ± 3 | 54 ± 3 | -16 ± 4 | -38 ± 4 | This study |
| Nipigon Bay (NB) | 36 ± 3 | 39 ± 4 | -19 ± 4 | 36 ± 4 | 0 ± 5 | -55 ± 6 | This study |
| Open Lake Superior (isothermal) | 58 ± 4 | 39 ± 4 | 2 ± 3 | 57 ± 4 | -1 ± 5 | -49 ± 5 | This study |
| Open Lake Superior (stratified) | 56 ± 4 | 38 ± 3 | 17 ± 4 | 58 ± 4 | -2 ± 5 | -41 ± 5 | This study |
| Pacific coastal ocean | 100 ± 4 | -200 | 100 | 76 ± 5 | 24 ± 6 | 24 ± 5 | Druffel and Williams 1991; Druffel and |

| | | | | | | | |
|---|---|---|---|---|---|---|---|
| | | | | | | | Williams 1990 |
| North Central Pacific (NCP) | 132 | -200 | 139 ± 9 | 124 ± 46 | 8 | 15 | Williams et al. 1987; Druffel and Williams 1990 |
| Northeast Pacific (NEP) | 155 | -200 | 82 | 86 ± 40 | 69 | -4 | Williams et al. 1987; Druffel et al. 1996 |

# CHAPTER SEVEN

# SUMMARY OF CONCLUSIONS AND FUTURE WORK

The sources, biogeochemical cycling and fate of organic carbon in Lake Superior were comprehensively examined using natural abundance radiocarbon and stable isotopic compositions of bulk DOC, POC, and DIC, size fractionated organic carbon, and biochemical compound-classes of high molecular weight DOC. Results from multiple years (2007-2010) at the western arm of the lake show seasonal variability (stratified vs. isothermal) and slight annual variations in the abundances and stable isotopic compositions of DIC, DOC, and POC driven mostly by biochemical processes such as photosynthesis and respiration in the lake.

The DIC reservoir appears to reset rapidly in the lake, with radiocarbon values that appear to record atmospheric radiocarbon levels from three years previous to sampling. Radiocarbon ($\Delta^{14}C$) values of DIC decreased from 74-84‰ in September 2007 to 52-57‰ in August 2010 due to equilibration and/or exchange with other carbon reservoirs. Similarly, $\Delta^{14}C$ of atmospheric $CO_2$ in the Lake's watershed decreased from 64‰ in June 2004 (Hsueh et al. 2007) to 36‰ in September 2010 due to equilibration of the bomb $^{14}C$ with carbon reservoirs in the biosphere. The offset between radiocarbon values in DIC and atmospheric $CO_2$ indicates that the DIC pool is not yet in equilibrium with atmospheric $^{14}CO_2$. There was no significant reservoir effect in the lake as the DIC pool was not $^{14}C$-depleted relative to the atmospheric $CO_2$, consistent with the lake's large surface area, the absence of significant limestone and dolomite deposits in its drainage basin, and the biannual vertical mixing of the lake water column. Distinct processes such as in-situ photosynthesis and sedimentary resuspension appear to be the primary regulators of the radiocarbon values of bulk POC in the lake water column. The bulk DOC pool persists longer than DIC with turnover times of up to 6 decades, whereas suspended bulk POC was the oldest OC pool in the lakes, most likely due to resuspension from the lake sediments.

The results from the lake-wide study show that DIC concentration was laterally homogeneous and tightly coupled to the physical/thermal regime of the lake. DOC and POC concentrations were laterally variable across the lake due to varying extent of photosynthesis and respiration. Offshore radiocarbon values of DIC exhibited lateral homogeneity and were more enriched than atmospheric $CO_2$. The radiocarbon values of nearshore DIC were similar to atmospheric $CO_2$. Bulk DOC was consistently modern throughout the lake indicating semi-lability with turnover time $\leq$ 60 years. Radiocarbon values of POC were spatially heterogeneous with old POC mostly seen in the deeper basins, and also in the deep waters during stratification, and attributed to resuspension.

Organic carbon size fractionation results show that LMW DOC was the largest pool of organic carbon in the lake, constituting 76-92% of DOC in the lake. LMW DOM in the lake was modern, semi-labile, with turnover time of $\leq$ 60 years. Coupled isotopic and elemental signatures indicate that HMW DOC is mostly from contemporary origin and recycles rapidly. NMR shows that HMW DOM has more aliphatic compounds and acetate, with only trace aromatic structures; this compound-class distribution indicates algal and microbial sources and possibly photochemically and microbially altered terrestrial material.

Radiocarbon study of the biochemical compound classes within the HMW DOM shows that total hydrolyzable carbohydrates and amino acids have modern (post-1950) $\Delta^{14}C$ values reflecting a recent photosynthetic origin. Solvent extractable lipid was relatively $^{13}C$-depleted and persisted much longer (up to 2,320 BP years) than the lake water residence time. Coupled $\Delta^{14}C$ and $\delta^{13}C$ values indicate multiple sources, and the co- existence of different formation pathways for the acid insoluble fraction within the HMW DOM in the lake.

Finally, the radiocarbon and stable isotopic composition of mesozooplankton shows that algal biomass from in situ primary production contributed a disproportionately large amount (mean 85%) of the carbon in zooplankton biomass in Lake Superior. Although terrestrial organic carbon and sedimentary organic carbon were significant portions of the available basal food resources, these contributed only trace amounts (mean $\leq$ 4%) to the zooplankton biomass. The fate of the old POC in the lake bears further investigation as it does not appear to be channeled up the lake food web.

To gain holistic insights into the OC budget and OC biogeochemistry, winter time data on primary production, respiration, abundances of DOC, POC, and DIC and their isotopic compositions would be needed. Also, to better understand the dynamics of bulk POC in the lake, the mean ages and potential sources of the sinking POC pool should be examined by analyzing the radiocarbon values and biomarker composition of sinking POC. Sinking particulate organic matter is the most dynamic of all the organic carbon pools and plays important roles in sedimentation and biogeochemical cycling of a variety of elements in the water column. A better constraint on the sources of sinking POC and processes that control it will enhance our understanding of the dynamics of benthic-pelagic coupling and the importance of sediment resuspension in nutrient recycling within the lake water column.

Although the radiocarbon values of mesozooplankton in the lake do not show incorporation of the old carbon into the zooplankton biomass, there is the need to examine the radiocarbon values of specific zooplankton species (benthic vs. pelagic feeders) and fishes in the lake in order to comprehensively assess if the old carbon from resuspension sustains higher trophic levels in the lake. Also, the radiocarbon composition of bacterial biomass, and also bacterially respired carbon in the lake should be examined to determine the carbon sources fueling microbial respiration and bacterial production in the lake water column, and to ultimately determine if old carbon is processed within the microbial loop. Finally, the amount, turnover times, and reactivity of riverine organic matter entering the lake should be assessed at a much higher spatial scale and temporal resolution to better constrain terrestrial influence on the lake carbon biogeochemistry.

# CHAPTER EIGHT

# REFERENCES


Abbott, M. B., and T. W. Stanfford 1996. Radiocarbon geochemistry of modern and ancient Arctic Lake systems, Baffin Island, Canada, *Quaternary Research*, *45*, 300-311.

Abdulla, H. A. N, E. C. Minor, and P. G. Hatcher. 2010b. Using two-dimensional correlations of $^{13}C$ NMR and FTIR to investigate changes in the chemical composition of dissolved organic matter along an estuarine transect. *Environ. Sci. Technol.* 44(21), 8044-8049.

Abdulla, H. A. N., E. C. Minor, R. F. Dias, and P. G. Hatcher. 2010. Changes in the compound classes of dissolved organic matter along an estuarine transect: A study using FTIR and $^{13}C$ NMR. *Geochim. Cosmochim. Acta* 74, 3815-3838.

Abell, J., S. Emerson, and P. Renaud. 2000. Distributions of TOP, TON and TOC in the North Pacific subtropical gyre: Implications for nutrient supply in the surface ocean and remineralization in the upper thermocline. *J. Mar. Res.* 58, 203-222.

Aitkenhead, J. A., and W. McDowell II. 2000. Soil C:N ratio as a predictor of annual riverine DOC flux at local and global scales. Global Biogeochem. Cycles **14**: 127-138.

Alin, S. R., and T. C. Johnson (2007), Carbon cycling in large lakes of the world: A synthesis of production, burial, and lake-atmosphere exchange estimates, *Global Biogeochem. Cycles*, *21*, GB3002, doi: 10.1029/2006GB002881.

Alin, S. R., R. Aalta, M. A. Goni, J. E. Richey. 2008. Biogeochemical characterization of carbon sources in the Strickland and Fly rivers, Papua New Guinea. J. Geophys. Res. 113, F01S05, doi:10.1029/2006JF000625.

Aller, R. C., Blair, N. E., and Brunskill, G. J.: Early diagenetic cycling, incineration, and burial of sedimentary organic carbon in the central Gulf of Papua (Papua New Guinea), J. Geophys. Res. 113, F01S09, doi:10.1029/2006JF000689, 2008.

Alperin, M. J., D. B. Albert, and C. S. Martens. 1994. Seasonal variations in production and consumption rates of dissolved organic carbon in an organic-rich coastal sediment. Geochim. Cosmochim. Acta **58:** 4909-4929.

Aluwihare, L. I., D. J. Repeta, and R. F. Chen. 1997. A major biopolymeric component to dissolved organic carbon in surface seawater. *Nature* 387, 166-169.

Aluwihare, L.I., D.J. Repeta, R.F. Chen. 2002. Chemical composition and cycling of dissolved organic matter in the Mid-Atlantic Bight. *Deep-Sea Research II* 49, 4421–4437.

Aluwihare, L.I., D.J. Repeta, R.F. Chen. 2002. Chemical composition and cycling of dissolved organic matter in the Mid-Atlantic Bight. *Deep-Sea Research II* 49, 4421–4437. America's largest river. Science **301:** 88-91.

Amon, M. W., and R. Benner. 1994. Rapid cycling of high molecular weight dissolved organic matter in the ocean. *Nature* 369, 549-552.

Amon, M. W., and R. Benner. 1996. Bacteria utilization of different size classes of dissolved organic matter. *Limnol. Oceanogr.* 41, 41-51.

Anderson, D. J., T. B. Bloem, and J. V. Higgins. 1998. Sub-sampling technique for the determination of particulate-phase organic carbon in water. J. Great Lakes Res. **24:** 838-844.

Arnosti, C. and M. Holmer. 2003. Carbon cycling in a continental margin sediment: contrasts between organic matter characteristics and remineralization rates and pathways. *Estuarine, Coastal and Shelf Science* 58, 197-208.

Arnosti, C., D. J. Repeta, and N. V. Blough. 1994. Rapid bacterial degradation of polysaccharides in anoxic marine systems. *Geochim. Cosmochim. Acta* 58, 2639-2652.

Austin, J. 2013. The potential for Autonomous Underwater Gliders in large lake research. *Journal of Great Lakes Research* **39**, Supplement 1, 8-13.

Austin, J. A., and Colman, S. M. 2007. Lake Superior summer water temperatures are increasing more rapidly than regional air temperatures: A positive ice-albedo feedback. *Geophysical Research Letters*, *34*(6).

Azam, F., and R. A. Long. 2001. Sea snow microcosms. *Nature* 414, 495-498.

Azam, F., T. Fenchel, J. G. Field, J. S. Gray, L. Meyer-Reil, and F. Thingstad. 1983. The cological role of water column microbes in the sea. *Mar. Ecol. Prog. Ser.* 10, 257-263.

Bacon, M. P., and R. F. Anderson. 1982. Distribution of thorium isotopes between dissolved and particulate forms in the deep sea. *J. Geophys. Res.* 87, 2045-2056.

Bahnick, D. A., T. P. Markee, C. A. Anderson, and R. K. Roubal (1978), Chemical loadings to southwestern Lake Superior from red clay erosion and resuspension, *J. Great Lakes Res.*, *4*, 186-193.

Baker, J. E., and S. J. Eisenreich. 1989. PCBs and PAHs as tracers of particulate dynamics in large lakes, *J. Great Lakes Research*, *15*, 84-103.

Baker, J. E., and S. J. Eisenreich. 1985. Chlorinated hydrocarbon cycling in the benthic nepheloid layer of Lake Superior. Environ. Sci. Technol **19:** 854-861.

Balakrishna, K., I.T. Kumar, G. Srinikethan, and G. Muderaya. 2006. Natural and anthropogenic factors controlling the dissolved organic carbon concentrations and fluxes in a large tropical river, India. *Environ. Monit. Assess*. 122, 355-364.

Balcer, M. D. 1984. *Zooplankton of the Great Lakes: A Guide to the identification and ecology of the common crustacean species.* University of Wisconsin Press, Ltd.

Baldock, J. A., C. A. Masiello, Y. Gelinas, J. I. Hedges. 2004. Cycling and composition of organic matter in terrestrial and marine ecosystems. *Marine Chemistry* 92, 39-64.

Banse, K. 1977. Determining the carbon-to-chlorophyll ratio of natural phytoplankton. Marine Biology **41:** 199-212.

Barber, R. T. 1968. Dissolved organic carbon from deep waters resists microbial oxidation. *Nature* 220, 274-275.

Barbiero, R. P. and Tuchman, M. L.2004. The deep chlorophyll maximum in Lake Superior. J. Great Lakes Res, 30, 256-268.

Bauer, J. E., and E. R. M. Druffel. 1998. Ocean margins as a significant source of organic matter to the deep open ocean, *Nature*, *392*, 482-485.

Bauer, J. E., E. R. M. Druffel, D. M. Wolgast, and S. Griffin. 2001. Sources and cycling of dissolved and particulate organic radiocarbon in the northwest Atlantic continental margin. *Global Biogeochem. Cycles* 15, 615-636.

Bauer, J. E., E. R. M. Druffel, D. M. Wolgast, and S. Griffin. 2002. Temporal and regional variability in sources and cycling of DOC and POC in the northwest Atlantic continental shelf and slope. Deep Sea Research II **49:** 387–4419.

Bauer, J.E. 2002. Carbon isotopic composition of DOM. In: D.A. Hansell and C.A. Carlson (Eds). *Biogeochemistry of Marine Dissolved Organic Matter*, Academic Press, Boston, p. 405 453.

Bauer, J.E. and Druffel, E.R.M. 1998. Ocean margins as a significant source of organic matter to the deep open ocean. *Nature* **392**, 482-485.

Bauer, J.E., Williams, P.M. and Druffel, E.R.M. 1992. $^{14}C$ activity of dissolved organic carbon fractions in the north-central Pacific and Sargasso Sea. *Nature* **357**, 667-670.

Beaupre, S. R., and E. R. M. Druffel. 2009. Constraining the propagation of bomb-radiocarbon through the dissolved organic carbon (DOC) pool in the northeast Pacific Ocean. *Deep-Sea Research I* 56: 1717-1726.

Beaupre, S. R., E. R. M. Druffel, and S. Griffin. 2007. A low-blank photochemical extraction system for concentration and isotopic analyses of marine dissolved organic carbon. *Limnol. Oceanogr. Methods* 5, 174-184.

Benner, R. 2004. What happens to terrestrial organic matter in the ocean? *Marine Chemistry* 92, 307-310.

Benner, R. and Kaiser, K. 2003. Abundance of amino sugars and peptidoglycan in marine particulate and dissolved organic matter. *Limnology and Oceanography* **48**, 118-128.

Benner, R., and J. I. Hedges. 1993. A test of the accuracy of fresh water DOC measurements by high-temperature catalytic oxidation and UV-promoted persulfate oxidation. *Marine Chemistry* 41, 161-165.

Benner, R., and S. Opsahl. 2001. Molecular indicators of the sources and transformations of dissolved organic matter in the Mississippi River plume. *Org. Geochem.* 32, 597-611.

Benner, R., B. Biddanda, B. Black, and M. McCarthy. 1997. Abundance, size distribution, and stable carbon and nitrogen isotopic compositions of marine organic matter isolated by tangential-flow ultrafiltration. *Marine Chemistry* 57, 243-263.

Benner, R., J. D. Pakulski, M. McCarthy, J. I. Hedges, and P. G. Hatcher. 1992. Bulk chemical characteristics of dissolved organic matter in the ocean. *Science* 255, 1561-1564.

Benner, R., S. Opsahl, G. Chin-Leo, J.E. Richey, B.R. Forsberg. 1995. Bacterial carbon metabolism in the Amazon river system. *Limnol. Oceanogr.* 40, 1262-1270.

Benner, R., S. Opsahl. 2001. Molecular indicators of the sources and transformations of dissolved organic matter in the Mississippi river plume. *Org. Geochem*. 32, 597-611.

Bennington, V. (2010), Carbon cycle variability of the North Atlantic Ocean and Lake Superior, Ph.D Thesis, University of Wisconsin-Madison.

Bennington, V., G. A. McKinley, N. Kimura, and C. H. Wu (2010), General circulation of Lake Superior: Mean, variability, and trends from 1979 to 2006, *J. Geophys. Res., 115*, C12015, doi: 10.1029/2010JC006261, 2010.

Bennington, V., McKinley, G.A., Urban, N.R. and McDonald, C.P. (2012)

Can spatial heterogeneity explain the perceived imbalance in Lake Superior's carbon budget? A model study. Journal of Geophysical Research: *Biogeosciences* **117**, G03020.

Berggren M., Strom, L., Laudon, H., Karlsson, J., Jonsson, A., Giesler, R., Bergstrom, A. K., and Jansson, M.: Lake Secondary production fueled by rapid transfer of low molecular weight organic carbon from terrestrial sources to aquatic consumers, Ecol. Lett. 13, 870-880, 2010.

Bianchi, T. S., J. E. Bauer, E. R. M. Druffel, and C. D. Lambert. 1998. Pyrophaeophorbide-a as a tracer of suspended particulate organic matter from NE Pacific continental margin. Deep Sea Res. II **45:** 715-731.

Bianchi, T. S., T. Filley, K. Dria, and P. G. Hatcher. 2004. Temporal variability in sources of dissolved organic carbon in the lower Mississippi River. *Geochim. Cosmochim. Acta* 68, 959-967.

Bianchi, T.S., C. Lambert, P.H. Santschi, M. Baskaran, and L. Guo. 1995. Plant pigments as biomarkers of high-molecular-weight dissolved organic matter. *Limnol. Oceanogr.* 40, 422-428.

Biddanda, B.A. and J. B. Cotner, 2003. Enhancement of dissolved organic matter bioavailability by sunlight and its role in the carbon cycle of Lakes Superior and Michigan. *J. Great Lakes Res*. 29, 228-241.

Boon, J. J., V. A. Klap, and T. I. Eglinton. 1998. Molecular characterization of microgram amounts of oceanic colloidal organic matter by direct temperature-resolved ammonia chemical ionization mass spectrometry. *Org. Geochem*. 29, 1051-1061.

Boyer, L. F., and J. Hedrick (1989), Submersible-deployed video sediment profile camera system for benthic studies, *J. Great Lakes Res*., *15* (1), 34-45.

Brett, M. T., Kainz, M. J., Taipale, S. J., and Seshan, H.: Phytoplankton, not allochthonous carbon, sustains herbivorous zooplankton production, PNAS 106, 21197-21201, 2009.

Buckley, D. R., K. J. Rockne, A. Li, and W. J. Mills (2004), Soot deposition in the Great Lakes: Implications for semi-volatile hydrophobic organic pollutant deposition, *Environ. Sci. Technol.*, *38*, 1732-1739.

Burdige, D. J. 2002. Sediment porewaters, p. 611-663. *In* D. A. Hansell and C. A. Carlson [eds.], Biogeochemistry of marine dissolved organic matter. Academic Press.

Burdige, D. J. 2007. Preservation of organic matter in marine sediments: Controls, mechanisms, and an imbalance in sediment organic carbon budgets?. Chem. Rev. **107:** 467-485.

Burkhardt, S., U. Riebesell, and I. Zondervan. 1999. Effects of growth rate, $CO_2$ concentration, and cell size on the stable carbon isotope fractionation in marine phytoplankton. Geochim. Cosmochim. Acta **63:** 3729-3741.

Bushaw, K. L., Zepp, R. G., Tarr, M. A., Schulz-Jander, D., Bourbonniere, R. A., Hodson, R. E., Miller, W. L., Bronk, D. A., and Moran, M. A. 1996. Photochemical release of biologically available nitrogen from aquatic dissolved organic matter. *Nature*, 381, 404-407.

Butman, D., P. Raymond, N. H. Oh, and K. Mull (2007), Quantity, $^{14}C$ age and lability of desorbed soil organic carbon in fresh water and seawater, *Org. Geochem.*, *38*, 1547-1557.

Caraco, N. F. 1995. Influence of human populations on phosphorus transfers to aquatic systems: a regional scale study using large rivers, p. 235-244. *In* H. Tiessen [ed.], Phosphorus in the Global Environment: Transfers, cycles and management. Wiley.

Caraco, N., Bauer, J. E., Cole, J. J., Petsch, S., and Raymond, P.: Millenial-aged organic carbon subsidies to a modern river food web, Ecology 91, 2385-2393, 2010.

Carlson, T. W. (1982), Deep-water currents and their effect on sedimentation in Lake Superior, Ph.D Thesis, University of Minnesota.

Carpenter, S. R., Cole, J. J., Pace, M. L., Van de Bogert, M., Bade, D. L., Bastviken, D., ... & Kritzberg, E. S. 2005. Ecosystem subsidies: terrestrial support of aquatic food webs from 13C addition to contrasting lakes. *Ecology*, *86*(10), 2737-2750.

Cauchie H. M. 2002. Chitin production by arthropods in the hydrosphere. *Hydrobiologia* 470, 63-69.

Cauwet, G. 2002. DOM in the coastal zone. *In* Biogeochemistry of marine dissolved organic matter (D. A. Hansell and C. A. Carlson, Eds.), pp 579-609. Academic Press, San Diego.

Cawley, K.M., Murray, A.E., Doran, P.T., Kenig, F., Stubbins, A., Chen, H., Hatcher, P.G. and McKnight, D.M. (2016) Characterization of dissolved organic material in the interstitial brine of Lake Vida, Antarctica. *Geochimica et Cosmochimica Acta* **183**, 63-78.

Cherrier, J., Bauer, J. E., Druffel, E. R. M., Coffin, R. B., and Chanton, J. P.: Radiocarbon in marine bacteria: Evidence for the ages of assimilated carbon, Limnol. Oceanogr. 44, 730-736, 1999.

Chin, Y.-P., Aiken, G. and O'Loughlin, E. (1994) Molecular weight, polydispersity, and spectroscopic properties of aquatic humic substances. *Environmental Science & Technology* **28**, 1853-1858.

Chiou, C.T., R. L. Malcolm, T. I. Brinton, and D. E. Kile. 1986. Water solubility enhancement of some organic pollutants and pesticides by dissolved humic and fulvic acids. *Environ. Sci. Tech.* 20, 502-508.

Churchill, J. H. and A. J. Williams (2004), Bottom stress generation and sediment transport over the shelf and slope off of Lake Superior's Keweenaw peninsula, *J. Geophys. Res.*, 109, C10S04, doi:10.1029/2003JC001997.

Cifuentes, L. and Salata, G. (2001) Significance of carbon isotope discrimination between bulk carbon and extracted phospholipid fatty acids in selected terrestrial and marine environments. *Organic Geochemistry* **32**, 613-621.

Cloern, J. E., E. A. Canuel, D. Harris. 2002. Stable carbon and nitrogen isotope composition of aquatic and terrestrial plants of the San Francisco Bay estuarine system. *Limnol. Oceanogr*. 47, 713-729.

Conde-Porcuna, J. M., Ramos-Rodriguez, E., and Perez-Martinez, C. 2002. Correlations between nutrient concentrations and zooplankton populations in a mesotrophic reservoir. *Freshwater Biology*, *47*(8), 1463-1473.

Cole J. J., Carpenter, S. R., Kitchell, J. F., and Pace, M. L.: Pathways of organic carbon utilization in small lakes: Results from a whole-lake C-13 addition and coupled model, Limnol Oceanogr. 47, 1664-1675, 2002.

Cole, J. J., and N. F. Caraco. 2001. Carbon in catchments: connecting terrestrial carbon losses with aquatic metabolism. Marine and Freshwater Research **52:** 101-110.

Cole, J. J., Carpenter, S. R., Pace, M. L., Van de Bogert, M. C., Kitchell, J. L., and Hodgson, J. R.:Differential support of lake food webs by three types of terrestrial organic carbon, Ecol. Lett. 9, 558-568, 2006.

Cole, J. J., N. F. Caraco, G. W. Kling, and T. K. Kratz (1994), Carbon dioxide supersaturation in the surface waters of lakes, *Science*, *256*, 1568-1570.

Cole, J. J., Y. T Prairie, N. F Caraco, W. H McDowell, L. J. Tranvik, R. G. Striegl, C. M. Duarte, P. Kortelainen, J. A. Downing, J. J. Middelburg, and J. Melack (2007), Plumbing the global carbon cycle: Integrating inland waters into the terrestrial carbon budget, *Ecosystem*, *10*, 171-184.

Cole, J. L., Carpenter, S. R., Kitchell, J., Pace , M. L., Solomon, C. T., and Weidel, B.: Strong evidence for terrestrial support of zooplankton in small lakes based on stable isotopes of carbon, nitrogen, and hydrogen, PNAS 108, 1975-1980, 2011.

Cole, J.J., Carpenter, S.R., Kitchell, J.F. and Pace, M.L. (2002) Pathways of organic carbon utilization in small lakes: Results from a whole-lake $^{13}C$ addition and coupled model. *Limnology and Oceanography* **47**, 1664-1675.

Cole, J.J., Cole, J.J., Caraco, N.F. and Caraco, N.F. 2001. Carbon in catchments: connecting terrestrial carbon losses with aquatic metabolism. *Marine and Freshwater Research* **52**, 101-110.

Cole, J.J., N.F. Caraco, G.W. Kling, T.K. Kratz. 1994. Carbon Dioxide Supersaturation in the Surface Waters of lakes. *Science* 256, 1568-1570.

Conde-Porcuna, J. M., Ramos-Rodriguez, E., and Perez-Martinez, C. 2002. Correlations between nutrient concentrations and zooplankton populations in a mesotrophic reservoir. *Freshwater Biology*, *47*(8), 1463-1473.

Cotner, J. B., B. A. Biddanda, W. Makino, and E. Stets. 2004. Organic carbon biogeochemistry of Lake Superior, *Aquatic Ecosystem Health and Management*, *7*, 451-464.

Cotner, J. B., Johengen, T. H., & Biddanda, B. A. 2000. Intense winter heterotrophic production stimulated by benthic resuspension. *Limnology and Oceanography*, *45*(7), 1672-1676.

Crow, S. E., K. Lajtha, T. R. Filley, C. W. Swanston, R. D. Bowden, and B. A. Caldwell. 2009. Sources of plant-derived carbon and stability of organic matter in soil: implications for global change. Global Change Biology **15:** 2003-2019.

Dai, M. H., J. M. Martin, and G. Cauwet. 1995. The significant role of colloids in the transport and transformation of organic carbon and associated trace metals (Cd, Cu, and Ni) in the Rhone delta (France). *Marine Chemistry* 51, 159-175.

de Kluijver, A., Schoon, P.L., Downing, J.A., Schouten, S. and Middelburg, J.J. (2014) Stable carbon isotope biogeochemistry of lakes along a trophic gradient. *Biogeosciences* **11**, 6265-6276.

Dean, W., and E. Gorham. 1998. Magnitude and significance of carbon burial in lakes, reservoirs, and peatlands, *Geology*, *26* (6), 535-538.

Del Giorgio, P. A., and France, R. L.: Ecosystem-specific patterns in the relationship between zooplankton and POM or microplankton $\delta^{13}C$, Limnol. Oceanogr. 41, 359-365, 1996.

del Giorgio, P.A. and Peters, R.H. (1994) Patterns in planktonic P:R ratios in lakes: Influence of lake trophy and dissolved organic carbon. *Limnology and Oceanography* **39**, 772-787.

DeMott, W. R., & Gulati, R. D. 1999. Phosphorus limitation in Daphnia: evidence from a long term study of three hypereutrophic Dutch lakes. *Limnology and Oceanography*, *44*(6), 1557-1564.

Deniro, M. J., and Epstein, S.: Influence of diet on the distribution of carbon isotopes in animals, Geochim. Cosmochim. Acta 42, 495-506, 1978.

DeNiro, M.J. and Epstein, S. (1977) Mechanism of carbon isotope fractionation associated with lipid synthesis. *Science* **197**, 261-263.

Dittmar, T., Koch, B., Hertkorn, N. and Kattner, G. (2008) A simple and efficient method for the solid-phase extraction of dissolved organic matter (SPE-DOM) from seawater. *Limnol. Oceanogr. Methods* **6**, 230-235.

Doering, P.H., C.A. Oviatt, J.H. McKenna, L.W. Reed. 1994. Mixing behavior of dissolved organic carbon and its potential biological significance in the Pawcatuck River estuary. *Estuaries* 17, 521-536.

Donahue, D. J., Linick, T. W., & Jull, A. T. 1990. Isotope-ratio and background corrections for accelerator mass spectrometry radiocarbon measurements. *Radiocarbon*, *32*(2), 135-142.

Druffel, E. R. M., and J. E. Bauer. 2000. Radiocarbon distribution in Southern Ocean dissolved and particulate organic matter. Geophys. Res. Lett. **27:** 1495-1498.

Druffel, E. R. M., and P. M. Williams. 1990. Identification of a deep marine source of particulate organic carbon using bomb $^{14}C$. Nature **347:** 172-174.

Druffel, E. R. M., and Williams, P. M.: Radiocarbon in seawater and organisms from the Pacific coast of Baja California. Radiocarbon 33, 291-296, 1991.

Druffel, E. R. M., Bauer, J. E., Williams , P. M., Griffin, S., and Wolgast, D.: Seasonal variability of particulate organic radiocarbon in the Northeast Pacific Ocean, J. Geophys. Res. 101, 543-552, 1996.

Druffel, E. R. M., J. E. Bauer, P. M. Williams, S. Griffin, and D. Wolgast. 1996. Seasonal variability of particulate organic radiocarbon in the northeast Pacific Ocean. J. Geophys. Res. **101:** 20543-20552.

Druffel, E. R. M., P. M. Williams J. E. Bauer, and J. R. Ertel, 1992. Cycling of dissolved and particulate organic matter in the open ocean. J. Geophys. Res. 97, 15,639-15,659.

Druffel, E. R. M., P. M. Williams, J. E. Bauer, and J. R. Ertel. 1992. Cycling of dissolved and particulate organic matter in the open ocean. *J. Geophys. Res*. 97, 15639-15659.

Druffel, E. R. M., S. Griffin, J. E. Bauer, D. M. Wolgast, and X. C. Wang. 1998. Distribution of particulate organic carbon and radiocarbon in the water column from the upper slope to the abyssal NE Pacific Ocean. Deep Sea Research II **45:** 667–687.

Druffel, E.R.M., and J.E. Bauer. 2000. Radiocarbon distribution in Southern Ocean dissolved and particulate organic matter. *Geophys. Res. Lett*. 27, 1495-1498.

Druffel, E.R.M., D. Zhang, X. Xu, L.A. Ziolkowski, J.R. Southon, G.M. Santos, and S.E. Trumbore. 2010. Compound specific radiocarbon analyses of phospholipid fatty acids and alkanes in ocean sediments. *Radiocarbon* 52, 1215-1223.

Eadie, B. J., D. J. Schwab, T. H. Johengen, P. J. Lavrentyev, G. S. Miller, R. E. Holland, G. A Leshkevich, M. B. Lansing, N. R. Morehead, J. A. Robbins, N. Hawley, D. N. Edgington, and P. L. Van Hoof (2002), Particle transport, nutrient cycling, and algal community structure associated with a major winter-spring sediment resuspension event in southern Lake Michigan, *J. Great Lakes Res*., *28*, 324-337.

Eglinton, T. I., B. C. Benitez-Nelson, A. Pearson, A. P. McNichol, J. E. Bauer, and E. R. M. Druffel (1997), Variability of radiocarbon ages of individual organic compounds from marine sediments, *Science*, *277*, 796-799.

Engelhaupt, E., and T. S. Bianchi. 2001. Sources and composition of high molecular weight dissolved organic carbon in a southern Louisiana tidal stream (Bayou Trepagnier). *Limnol. Oceanogr.* 46, 917-926.

Environment Canada and The US EPA, State of the great lakes 2009, Zooplankton populations, pp. 97-101, 2009.

Falkowski, P. G. 1991. Species variability in the fractionation of $^{13}C$ and $^{12}C$ marine phytoplankton. J. Plankton Res. **13:** 21-28.

Findlay, S., R.L. Sinsabaugh, D.T. Fisher, and P. Franchini. 1998. Sources of dissolved organic carbon supporting planktonic bacterial production in the tidal freshwater Hudson River. *Ecosystems* 1, 227–239.

Finlay, J. C., R. W. Sterner, and S. Kumar (2007), Isotopic evidence for in-lake production of accumulating nitrate in Lake Superior, *Ecol. Appl.*, *17* (8), 2323–2332.

Fisher, T. R., J. D. Hagy, and E. Rochell-Newall. 1998. Dissolved and particulate organic carbon in Chesapeake Bay. *Estuaries* 21, 215-219.

Flood, R. D. (1989), Submersible studies of current-modified bottom topography in Lake Superior, *J. Great Lakes Res*., 15, 3-14.

Flood, R. D., and Johnson, T. C.: Side-scan sonar targets in Lake Superior-evidence for bedforms and sediment transport, *Sedimentology*, *31*, 311-333, 1984.

Fogel, M. L., & Cifuentes, L. A. 1993. Isotope fractionation during primary production. In *Organic geochemistry: principles and applications* (pp. 73-98). Boston, MA: Springer US.

Fox, L. E. 1983. The removal of dissolved humic acid during estuarine mixing. *Estuarine Coastal and Shelf Science* 16, 431-440.

Freeman, K. H., and J. M. Hayes. 1992. Fractionation of carbon isotopes by phytoplankton and estimates of ancient $CO_2$ levels. Global Biogeochem. Cycles. **6:** 185-198.

Fry, B. (1991) Stable isotope diagrams of freshwater food webs. *Ecology* **72**, 2293-2297.

Fry, B., and Sherr, E.: $\delta^{13}C$ measurements as indicators of carbon flow in marine and freshwater ecosystems, Contributions in Marine Science 27, 13-47, 1984.

Fry, B.: Stable isotopes diagrams of freshwater food webs. Ecology 72, 2293-2297, 1991.

Gannon, J. E., and Stemberger, R. S. (1978) Zooplankton (especially crustaceans and rotifers) as indicators of water quality. Transactions of the American Microscopical Society, 16-35.

Gogou, A., and D. J. Repeta (2010) Particluate-dissolved transformations as a sink for semi-labile dissolved organic matter: Chemical characterization of high molecular weight dissolved and surface-active organic matter in seawater and in diatom cultures. *Marine Chemistry* 121, 215-223.

Goni, M. A., K. C. Ruttenberg, and T. I. Eglington (1997), Sources and contribution of terrigenous organic carbon to surface sediments in the Gulf of Mexico, *Nature*, *389*, 275-278.

Goni, M. A., Monacci, N., Gisewhite, R., Ogston, A., Crockett, J., and Nittrouer, C. A.: Distribution and sources of particulate organic matter in the water column and sediments of the Fly River Delta, Gulf of Papua (Papua New Guinea). Estuarine Coastal and Shelf Science 69, 225-245, 2006.

Graven, H.D., Guilderson, T.P. and Keeling, R.F. (2012) Observations of radiocarbon in $CO_2$ at seven global sampling sites in the Scripps flask network: Analysis of spatial gradients and seasonal cycles. *Journal of Geophysical Research: Atmospheres* **117**, D02303.

Grey, J. R., Jones, R. I., and Sleep, D.: Seasonal changes in the importance of the source of the organic matter to the diet of zooplankton in Loch Ness, as indicated by stable isotope analysis. Limnol. Oceanogr. 46, 505-513, 2001.

Guildford, S. J., and R. E. Hecky (2000), Total nitrogen, phosphorus, and nutrient limitation in lakes and oceans: Is there a common relationship?, *Limnol. Oceanogr.*, *45*, 1213-1223.

Guo, L., and P. H. Santschi. 1996. A critical evaluation of the cross-flow ultrafiltration technique for sampling colloidal organic carbon in seawater. *Marine Chemistry* 55, 113-127.

Guo, L., and R. W. MacDonald. 2006. Sources and transport of terrigenous organic matter in the upper Yukon River: Evidence from isotopic ($\Delta^{14}C$, $\delta^{13}C$, $\delta^{15}C$) composition of dissolved, colloidal, and particulate phases. *Global Biogeochem. Cycles* 20, GB2011, doi:10.1029/2005GB002593.

Guo, L., C.H. Coleman and P.H. Santschi. 1994. The distribution of Colloidal and dissolved organic carbon in the Gulf of Mexico. *Marine Chemistry* 45, 105-119.

Guo, L., D. M. White, Chen Xu, P. H. Santschi. 2009. Chemical and isotopic composition of high molecular weight dissolved organic matter from the Mississippi River Plume. *Marine Chemistry* 114, 63-71.

Guo, L., Lehner, J.K., White, D.M. and Garland, D.S. (2003) Heterogeneity of natural organic matter from the Chena River, Alaska. *Water Research* **37**, 1015-1022.

Guo, L., P. H. Santschi, and K. Warnken. 1995. Dynamics of dissolved organic carbon (DOC) in oceanic environments. *Limnol. Oceanogr.* 40, 1392-1403.

Guo, L., P. H. Santschi, L. A. Cifuentes, S. E. Trumbore, and J. Southon. 1996. Cycling of high molecular weight dissolved organic matter in the Middle Atlantic Bight as revealed by carbon isotopic ($^{13}C$ and $^{14}C$) signatures. *Limnol. Oceanogr.* 41, 1242-1252.

Guo, L., P.H. Santschi, and K. Warnken. 1995. Dynamics of dissolved organic carbon (DOC) in oceanic environments. *Limnol. Oceanogr.* 40, 1392-1403.

Guo, L., Santschi, P.H., Cifuentes, L.A., Trumbore, S.E. and Southon, J. (1996) Cycling of high-molecular-weight dissolved organic matter in the Middle Atlantic Bight as revealed by carbon isotopic ($^{13}C$ and $^{14}C$) signatures. *Limnology and Oceanography* **41**, 1242-1252.

Gustafsson, O., F. Haghseta, C. Chan, J. MacFarlane, and P. M. Gschwend (1997), Quantification of the dilute sedimentary soot phase: Implications for PAH speciation and bioavailability, *Environ. Sci. Technol*, *31*, 203-209.

Gustafsson, O., T. D. Bucheli, Z. Kukulska, M. Andersson, C. Largeau, J. N. Rouzaud, C. M. Reddy, and T. I. Eglinton (2001), Evaluation of a protocol for the quantification of black carbon in sediments, *Global Biogeochem. Cycles*, *15*, 881-890.

Halfon, E. 1984.The composition of particulate organic matter in the euphotic zone of Lake Superior. J. Great Lakes Res. **10:** 299-306.

Hama, T., K. Yanagi, J. Hama. 2004. Decrease in molecular weight of photosynthetic products of marine phytoplankton during early diagenesis. *Limnol. Oceanogr*. 49, 471-481.

Hamilton, S. K., Sippel, S. J., and Bunn, S. E.: Separation of algae from detritus for stable isotope or ecological stoichiometry studies using density fractionation in colloidal silica, Limnol. Oceanogr. Methods 3, 149-157, 2005.

Hansell, D. A. and C. A. Carlson. 2002. Biogeochemistry of Marine Dissolved Organic Matter. Academic Press, San Diego.

Hansell, D. A., and C. A. Carlson. 1998. Deep-Ocean gradients in the concentration of dissolved organic carbon. *Nature* 395, 263-266.

Harvey, C. J., and J. F. Kitchell (2000), A stable isotope evaluation of the structure and spatial heterogeneity of a Lake Superior food web, *Can. J. Fish. Aquat. Sci.*, *57*, 1395-1403.

Hawley, N. (2000), Sediment resuspension near the Keweenaw Peninsula, Lake Superior during the fall and winter 1990-1991, *J. Great Lakes Res.*, *26*(4), 495-505.

Hayes, J. (1993) Factors controlling $^{13}C$ contents of sedimentary organic compounds: principles and evidence. *Marine Geology* **113**, 111-125.

Hayes, J. M. 2001. Fractionation of the isotopes of carbon and hydrogen in biosynthetic processes, p. 225-278. *In* J. W. Valley and D. R. Cole [eds.], Stable isotope geochemistry.

Hecky, R. E., P. Campbell, and L. L. Hendzel. 1993. The stoichiometry of carbon, nitrogen, and phosphorus in particulate matter of lakes and oceans. Limnol. Oceanogr. **38:** 709-724.

Hedges, J. 1992. Global biogeochemical cycles: progress and problems. *Marine Chemistry* 39, 67-93.

Hedges, J. I. and OTHERS. 1994. Origins and processing of organic matter in the Amazon River as indicated by carbohydrates and amino acids. *Limnol. Oceanogr.* 39, 743-761.

Hedges, J. I., G. Eglinton, P. G. Hatcher, D. L. Kirchman, C. Arnosti, S. Derenne, R. P. Evershed, K. Kogel-Knabner, J. W. de Leeuw, R. Littke, W. Michaelis, J. Rullkotter. 2000. The molecularly-uncharacterized component of non-living organic matter in natural environments. *Org. Geochem*. 31, 945-958.

Hedges, J. I., J.R. Ertel, P.D. Quay, P.M. Grootes, J.E. Richey, A.H. Devol, G.W. Farwell, F.W. Schmidt, and E. Salati. 1986. Organic carbon-14 in the Amazon River system. *Science* 231, 1129-1131.

Hedges, J. I., W. A. Clark, G. L. Cowie. 1988. Organic matter sources to the water column and surficial sediments of a marine bay. *Limnology and Oceanography* 33, 1116-1136.

Hedges, J.I. (1992) Global biogeochemical cycles: progress and problems. *Marine Chemistry* **39**, 67-93.

Hedges, J.I., P.G. Hatcher, J.E. Ertel, and K.J. Meyers-Schulte. 1992. A comparison of dissolved humic substances from seawater with Amazon River counterpart by $^{13}$CNMR spectrometry. *Geochim. Cosmochim. Acta*. 56, 1753-1757.

Hedges, J.I., R.G. Keil, and R. Benner. 1997. What happens to terrestrial organic matter in the ocean? *Org. Geochem*. 27, 195-212.

Helms, J.R., A. Stubbins, J.D. Ritchie, E.C. Minor, D.J. Kieber, and K. Mopper. 2008. Absorbance spectra slopes and slope ratios as indicators of molecular weight and sources of chromophoric dissolved organic matter. *Limnol. Oceanogr.* 53, 955-969.

Herbland, A., and A. Le Bouteiler. 1981. The size distribution of phytoplankton and particulate organic matter in the equatorial Atlantic Ocean: importance of ultraseston and consequences. J. Plankton Res. **3:** 659-673.

Herdendorf, C. E.: Distribution of the world's large lakes, p. 3-38. In M. M. Tilzer, and C. Serruya [ed.], Large Lakes. Springer-Verlag, 1990.

Hernes, P.J. and R. Benner. 2002. Transport and diagenesis of dissolved and particulate terrigenous organic matter in the North Pacific Ocean. *Deep-See Research I* 49, 2119-2132.

Hertkorn, N., R. Benner, M. Frommberger, P. Schmitt-Kopplin, M. Witt, K. Kaiser, A. Kettrup, and J. I. Hedges. 2006. Characterization of a major refractory component of marine dissolved organic matter. *Geochim. Cosmochim. Acta*, *70*(12), 2990-3010.

Hessen, D. O., Anderson, T., and Lyche, A.: Carbon metabolism in a humic lake: pool sizes and cycling through zooplankton, Limnol. Oceanogr. 35, 84-99, 1990.

Hessen, D. O.: Dissolved organic carbon in a humic lake: effects on bacteria production and respiration. Hydrobiologia 229, 115-123, 1992.

Hesshaimer, V., M. Heimann, and I. Levin (1994), Radiocarbon evidence for a smaller oceanic carbon dioxide sink than previously believed, *Nature*, *370*, 201-203.

Hoffman, J. C., G. S. Peterson, A. M. Cotter, and J. R. Kelly. 2010. Using stable isotope mixing in a Great Lakes coastal tributary to determine food web linkages in young fishes. Estuaries and Coasts **33:** 1391-1405.

Hopmans, E. C., J. W. H. Weijers, E. Schefuss, L. Herfort, J. S. S. Damste, and S. Schouten. 2004. A novel proxy for terrestrial organic matter in sediments based on branched and isoprenoid tetraether lipids. Earth Planet. Sci. Let. **224**:107-116.

Horvath, R. S.: Microbial co-metabolism and the degradation of organic compounds in nature. Bacteriol Rev. 36, 146-155, 1972.

Hsueh, D. Y., N. Y. Krakauer, J. T. Randerson, X. Xu, S. E. Trumbore, and J. R. Southon. 2007. Regional patterns of radiocarbon and fossil fuel-derived $CO_2$ in the surface air across North America. Geophys. Res. Lett. **34:** L02816, doi: 10.1029/2006GL027032

Hsueh, D.Y., Krakauer, N.Y., Randerson, J.T., Xu, X., Trumbore, S.E. and Southon, J.R. (2007) Regional patterns of radiocarbon and fossil fuel-derived CO2 in surface air across North America. *Geophysical Research Letters* **34**, L02816.

Hughen, K., S. Lehman, J. Southon, J. Overpeck, O. Marchal, C. Herring and J. Turnbull (2004), $^{14}C$ activity and global carbon cycle changes over the past 50,000 years, *Science*, *303*, 202-207.

Hwang, J. and Druffel, E.R. (2003) Lipid-like material as the source of the uncharacterized organic carbon in the ocean? *Science* **299**, 881-884.

Hwang, J. and Druffel, E.R.M. (2006) Carbon isotope ratios of organic compound fractions in oceanic suspended particles. *Geophysical Research Letters* **33**, L23610.

Hwang, J., and E. R. M. Druffel. 2003. Lipid-like material as the source of the uncharacterized organic carbon in the ocean. Science **299:** 881-884.

Hwang, J., and E. R. M. Druffel. 2006. Carbon isotope ratios of organic compound fractions in oceanic suspended particles. *Geophys. Res. Lett.* 33, L23610, doi: 10.1029/2006GL027928.

Hwang, J., Druffel, E.R., Eglinton, T.I. and Repeta, D.J. (2006) Source (s) and cycling of the nonhydrolyzable organic fraction of oceanic particles. *Geochimica et cosmochimica acta* **70**, 5162-5168.

Hwang, J., E. R. M. Druffel, and T. I. Eglinton (2010), Widespread influence of resuspended sediments on oceanic particulate organic carbon: Insights from radiocarbon and aluminum contents in sinking particles, *Global Biogeochem. Cycles*, *24*, GB4016, doi:10.1029/2010GB003802.

Hwang, J., E. R. M. Druffel, T. I. Eglinton, D. J. Repeta. 2006. Sources and cycling of the nonhydrolyzable organic fraction of oceanic particles. *Geochim. Cosmochim. Acta* 70, 5162-5168.

Hwang, J., E.R.M Druffel, and S. Griffin. 2004. Temporal variability of $\Delta^{14}C$, $\delta^{13}C$, and C/N in sinking particulate organic matter at the deep time series station in the northeast Pacific Ocean. *Global Biogeochem. Cycles*, 18, GB4015, doi:10.1029/2004GB002221.

Isao, K., Hara, S., Terauchi, K. and Kogure, K. (1990) Role of sub-micrometer particles in the ocean. *Nature* **345**, 242-244.

Ishikawa, N. F., Uchida, M., Shibata, Y., and Tayasu, I.: A new application of radiocarbon ($^{14}C$) concentrations to stream food web analysis, Nuclear Instruments and Methods in Physics Research B 268, 1175–1178, 2010.

Pace, M. L., Carpenter, S. R., Cole, J. J., Coloso, J. J., Kitchell, J. F., Hodgson, J. R., ... & Weidel, B. C. (2007). Does terrestrial organic carbon subsidize the planktonic food web in a clear-water lake?. *Limnology and Oceanography*, *52*(5), 2177-2189.

Jiao, N., G. J. Herndl, D A. Hansell, R. Benner, G. Kattner, S. W. Wilhelm, D. L. Kirchman, M. G. Weinbauer, T. Luo, F. Chen, and F. Azam. 2010. Microbial production of recalcitrant dissolved organic matter: long-term carbon storage in the global ocean. Nature Reviews Microbiology **8:** 593-599.

Johengen, T. H., B. A. Biddanda, and J. B. Cotner (2008), Stimulation of Lake Michigan plankton metabolism by sediment resuspension and river run-off, *J. Great Lakes Res*., *34*, 213-227.

Johnson, B. L., and C. A. Johnston (1995), Relationship of lithology and geomorphology to erosion of western Lake Superior coast, *J. Great Lakes Res*., *21*(1), 3-16.

Johnson, T. C., J. D. Halfman, W. H. Busch, and R. D. Flood (1984), Effects of bottom currents and fish on sedimentation in a deep-water, acustrine environment, *Geol. Soc. Amer. Bul*., *95*, 1425-1436.

Jones, R. I., and Salonen, K.: The importance of bacteria utilization of released phytoplankton photosynthate in two humic forest lakes in southern Finland, Holarctic Ecology 8, 133-140, 1985.

Jones, R. I., Grey, J., Sleep, D., and Quarmby, C.: An assessment, using stable isotopes, of the importance of allochthonous organic carbon sources to the pelagic food web in Loch Ness. Proc. R. Soc. Lond. B. 265, 105-111, 1998.

Jones, R. I.: The influence of humic substances on lacustrine planktonic food chains, Hydrobiologia 229, 73-91, 1992.

Jull, A. T., Burr, G. S., Beck, J. W., Hodgins, G. W., Biddulph, D. L., McHargue, L. R., & Lange, T. E. 2008. Accelerator mass spectrometry of long-lived light radionuclides. *Radioactivity in the Environment*, *11*, 241-262.

Kaiser, E., D.B. Arscott, K. Tockner, and B. Sulzberger. 2004. Sources and distribution of organic carbon and nitrogen in the Tagliamento River, Italy. *Aquatic Sciences* 66, 103-116.

Kaiser, K., and R. Benner. 2009. Biochemical composition and size distribution or organic matter at the Pacific and Atlantic time-Series stations. *Marine Chemistry* 113, 63-77.

Karlsson, J., A. Jonsson, M. Meili, and M. Jansson. 2003. Control of zooplankton dependence on allochthonous organic carbon in humic and clear-water lakes in northern Sweden. Limnol. Oceanogr. **48:** 269-276.

Karlsson, J., and Sawstrom, C. 2009. Benthic algae support zooplankton growth during winter in a clear-water lake, Oikos 118, 539-544.

Karlsson, J., Berggren, M., Ask, J., Byström, P., Jonsson, A., Laudon, H., & Jansson, M. 2012. Terrestrial organic matter support of lake food webs: Evidence from lake metabolism and stable hydrogen isotopes of consumers. *Limnology and Oceanography*, *57*(4), 1042-1048.

Karlsson, J., Jansson, M. and Jonsson, A. (2007) Respiration of allochthonous organic carbon in unproductive forest lakes determined by the Keeling plot method. *Limnology and Oceanography* **52**, 603-608.

Karlsson, J., Jonsson, A., Meili, M., and Jansson, M.: $\delta^{15}N$ of zooplankton species in subarctic lakes in northern Sweden: effects of diet and trophic fractionation, Freshwater Biology, 49, 526-534, 2004.

Karlsson, J., Jonsson, A., Meili, M., Jansson, M. Control of zooplankton dependence on allochthonous organic carbon in humic and clear-water lakes in northern Sweden, Limnol. Oceanogr. 48, 269-276, 2003.

Karlsson, J., M. Jansson, and A. Jonsson. 2007. Respiration of allochthonous organic carbon in unproductive forest lakes determined by the keeling plot method. *Limnol. Oceanogr.* 52, 603-608.

Karlsson, J.: Different carbon support for respiration and secondary production in unproductive lakes, Oikos 116, 1691-1696, 2007.

Keil R. G., D.B. Montluçon, F.G. Prahl, and J. I. Hedges, 1994. Sorptive preservation of labile organic matter in marine sediments. *Nature* 370, 549–551.

Kelly, C. A., E. Fee, P. S. Ramlal, J. W. M. Rudd, R. H. Hesslein, C. Anema, and E. U. Schindler (2001), Natural variability of carbon dioxide and net epilimnetic production in the surface waters of boreal lakes of different sizes, *Limnol. Oceanogr.*, *46*(5), 1054-1064.

Kemp, A. L. W., C. I. Dell, N. S. Harper (1978), Sedimentation rates and a sediment budget for Lake Superior, *J. Great Lakes Res.*, 4, 276-287.

Keough, J. R., C. A. Hagley, E. Ruzycki, and M. Sierszen (1998), $\delta^{13}C$ composition of primary producers and role of detritus in a freshwater coastal ecosystem, *Limnol. Oceanogr.*, *43*, 734-740.

Keough, J. R., M. E. Sierszen, and C. A. Hagley (1996), Analysis of a Lake Superior coastal food web with stable isotope techniques, *Limnol. Oceanogr.*, *41*, 136-146.

Kieber, R. J., R. F. Whitehead, and S. A. Skrabal (2006), Photochemical production of dissolved organic carbon from resuspended sediments, *Limnol. Oceonogr.*, *51*(5), 2187-2195.

Kiljunen, M., Grey, J., Sinisalo, T., Harrod, C., Immonen, H., and Jones, R. I.: A revised model for lipid-normalizing $\delta^{13}C$ values from aquatic organisms, with implications for isotopic mixing models, J. Appl. Ecol. 43, 1213-1222, 2006.

Kim, S., A.J. Simpson, E.B. Kujawinski, M.A. Freitas, P.G. Hatcher. 2003. High resolution Electrospray ionization mass spectrometry and 2D solution NMR for the analysis of DOM extracted by $C_{18}$ solid-phase disk. *Org. Geochem*. 34, 1325-1335.

Kirchman D. L. and White J. 1999. Hydrolysis and mineralization of chitin in the Delaware 629 Estuary. *Aquat. Microb. Ecol.* **18**, 187-196.

Klaminder, J., R. Bindler, H. Laudon, K. Bishop, O. Emteryd, and I. Renberg. 2006. Flux rates of atmospheric lead pollution within soils of a small catchment in Northern Sweden and their implications for future water quality. *Environ. Sci. Technol.* 40, 4639-4645.

Kling, G. W., Fry, B., and O'Brien, W. J.: Stable isotopes and planktonic trophic structure in arctic lakes, Ecology 73, 561-566, 1992.

Koarashi, J., M. Atarashi-Andoh, S. Ishizuka, S. Miura, T. Saito, and K. Hirai. 2009. Quantitative aspects of heterogeneity in soil organic matter dynamics in a cool-temperate Japanese beech forest: a radiocarbon-based approach. Global Change Biology **15**: 631-642.

Koelmans, A. A., and L. Prevo (2003), Production of dissolved organic carbon in aquatic sediment suspensions, *Water Res.*, *37*, 2217-2222.

Komada, T., and C. E. Reimers (2001), Resuspension induced partitioning of organic carbon between solid and solution phases from a river-ocean transition, *Mar. Chem.*, *76*, 155-174.

Komada, T., Druffel, E.R.M. and Hwang, J. (2005) Sedimentary rocks as sources of ancient organic carbon to the ocean: An investigation through $\Delta^{14}C$ and $\delta^{13}C$ signatures of organic compound classes. *Global Biogeochemical Cycles* **19**, GB2017.

Koprivnjak, J. F., P. H. Pfromm, E. Ingall, T. A. Vetter, P. Schmitt-Kopplin, N. Hertkorn, M. Frommberger, H. Knicker, and E. M. Perdue. 2009. Chemical and spectroscopic characterization of marine dissolved organic matter isolated using coupled reverse osmosis-electrodialysis. *Geochim. Comsochim. Acta* 73, 4215-4231.

Kritzberg, E. S., J. J. Cole, M. M. Pace, and W. Graneli. 2005. Bacterial growth on allochthonous carbon in humic and nutrient-enriched lakes: Results from whole-lake $^{13}C$ addition experiments. Ecosystems **9:** 489-499.

Kritzberg, E.S., J.J. Cole, M.M. Pace, and W. Graneli. 2005. Bacterial Growth on Allochthonous Carbon in Humic and Nutrient-enriched Lakes: Results from Whole-Lake $^{13}C$ Addition Experiments. *Ecosystems* 9, 489-499.

Kruger, B.R. (2014) Sources, cycling, and fate of organic matter in large lakes: insights from stable isotope and radiocarbon analysis in Lakes Malawi and Superior. University of Minnesota.

Kruger, B.R., Dalzell, B.J. and Minor, E.C. (2011) Effect of organic matter source and salinity on dissolved organic matter isolation via ultrafiltration and solid phase extraction. *Aquatic sciences* **73**, 405-417.

Kumar, S., J. C. Finlay, and W. R. Sterner (2011), Isotopic composition of nitrogen in suspended particulate matter of Lake Superior: implications for nutrient cycling and organic matter transformation, *Biogeochemistry*, *103*, 1-14.

Kumar, S., R. W. Sterner, and J. C. Finlay (2008), Nitrogen and carbon uptake dynamics in Lake Superior, *J. Geophys. Res.*, *113*, G04003. doi:10.1029/2008JG000720.

Kumara, S., Khan, J. A., & Thakur, M. S. 2018. Study on phytoplankton, zooplankton and ichthyo fauna of Motia lake. *Research & Reviews: Journal of Zoological Sciences*, *6*(2), 25-49.

Lalonde, K., Vähätalo, A. and Gélinas, Y. (2014) Revisiting the disappearance of terrestrial dissolved organic matter in the ocean: a $\delta^{13}C$ study. *Biogeosciences* **11**, 3707-3719.

Lam, B., A. Baer, M. Alaee, B. Lefebvre, A. Moser, A. Williams, and A. J. Simpson. 2007. Major structural components in freshwater dissolved organic matter. *Environ. Sci. Technol. 41*(24), 8240-8247.

Lam, B., and A. J. Simpson. 2008. Direct $^{1}H$ NMR spectroscopy of dissolved organic matter in natural waters. *Analyst* 133, 263-269.

Lam, D. C. (1978), Simulation of water circulation and chloride transports in Lake Superior for summer 1973, *J. Great Lakes Res.*, *4*, 343-349.

Lara, R.J., Rachold, V., Kattner, G., Hubberten, H.W., Guggenberger, G., Skoog, A. and Thomas, D.N. (1998) Dissolved organic matter and nutrients in the Lena River, Siberian Arctic: Characteristics and distribution. *Marine Chemistry* **59**, 301-309.

Laws, E. A., and R. R. Bidigare. 1997. Effect of growth rate and CO, concentration on carbon isotopic fractionation by the marine diatom Phaeodactylum tricornutum. Limnol. Oceanogr. **42:** 1552-1560.

Lee, C., S. Wakeham, C. Arnosti. 2004. Particulate organic matter in the sea: the composition conundrum. *AMBIO* 33, 565-575.

Leenheer, J. A.; Nanny, M. A.; McIntyre, C. 2003. Terpenoids as major precursors of dissolved organic matter in landfill leachates, surface water, and groundwater. *Environ. Sci. Technol. 37* (11), 2323–2331.

Lennon J. T., and Pfaff, L. E.: Source and supply of terrestrial organic matter affects aquatic microbial metabolism. Aquat. Microb. Ecol. 39, 107-119, 2005.

Levine, A. D., G. Tchobanoglous, and T. Asano. 1985. Characterization of the size distribution of contaminants in wastewater: Treatment and reuse implications. *J. Water Pollut. Control. Fed.* 57, 805-816.

Li, H., Minor, E.C. and Zigah, P.K. (2013) Diagenetic changes in Lake Superior sediments as seen from FTIR and 2D correlation spectroscopy. *Organic Geochemistry* **58**, 125-136.

Lobbes, J.M., Fitznar, H.P. and Kattner, G. (2000) Biogeochemical characteristics of dissolved and particulate organic matter in Russian rivers entering the Arctic Ocean. *Geochimica et Cosmochimica Acta* **64**, 2973-2983.

Loh, A. N., J. E. Bauer, and E. A. Canuel. 2006. Dissolved and particulate organic matter source-age characterization in the upper and lower Chesapeake Bay: A combined isotope and biochemical approach. *Limnol. Oceanogr.* 51, 1421-1431.

Loh, A.N., Bauer, J.E. and Druffel, E.R. 2004. Variable ageing and storage of dissolved organic components in the open ocean. *Nature* **430**, 877-881.

Lohrenz, S. E., G. L. Fahnenstiel, D. F. Millie, O. M. E. Schofield, T. Johengen, and T. Bergman (2004), Spring phytoplankton photosynthesis, growth, and primary production and relationships to a recurrent coastal sediment plume and river inputs in southeastern Lake Michigan, *J. Geophys. Res.*, *109*, C10S14, doi:10.1029/2004JC002383.

Lowell, T. V., G. J. Larson, J. D. Hughes, and G. H. Denton (1999), Age verification of the Lake Gribben forest bed and the younger dryas advance of the Laurentide ice sheet, *Canadian Journal of Earth Sciences*, *36*, 383-393.

M., Hodgson, J. R., Kitchell, J. F., and Kritzberg, E. S. 2005. Ecosystem subsidies: Terrestrial support of aquatic food webs from $^{13}C$ addition to contrasting lakes, Ecology, 86, 2737–2750.

Ma, X. and Green, S.A. (2004) Photochemical transformation of dissolved organic carbon in Lake Superior—an in-situ experiment. *Journal of Great Lakes Research* **30**, 97-112.

Ma, X., and S. A. Green. 2004. Photochemical transformation of dissolved organic carbon in Lake Supeior - an in-situ experiment. J. Great Lakes Res. **30:** 97-112.

Maier, W. J., and W. R. Swain .1978. Lake Superior organic carbon budget, *Water Res.*, *12*, 403-412.

Malcolm, R.L., 1990. The uniqueness of humic substances in each of soil, stream and marine environments. *Anal. Chim. Acta* 232, 19-30.

Mannino, A., and H. R. Harvey. 2000. Biochemical composition of particles and dissolved organic matter along an estuarine gradient: Sources and implications for DOM reactivity. *Limnol. Oceanogr*. 45, 775-788.

Marty, J., and D. Planas. 2008. Comparison of methods to determine algal $\delta^{13}C$ in freshwater. Limnol. Oceanogr.: Methods. **6:** 51-63.

Marwick, T.R., Tamooh, F., Teodoru, C.R., Borges, A.V., Darchambeau, F. and Bouillon, S. (2015) The age of river-transported carbon: A global perspective. *Global Biogeochemical Cycles*, 2014GB004911.

Masiello, C. A., and E. R. M. Druffel. 1998. Black carbon in deep-sea sediments. Science **280:** 1911-1913.

Matthews, B., and Mazumder, A. 2006. Habitat specialization and the exploitation of allochthonous carbon by zooplankton, Ecology 87, 2800-2812.

Matthews, B., and Mazumder, A. 2005. Temporal variation in body composition (C:N) helps explain seasonal patterns of zooplankton $\delta^{13}C$. Freshwater Biology 50, 502-515.

Mayer, L. M., L. L. Schick, and K. Skorko. 2006. Photodissolution of particulate organic matter from sediments, *Limnol. Oceanogr.*, *51*, 1064-1071.

Mayer, L. M., L. L. Schick., T. S. Bianchi, and L. A. Wysocki. 2009. Photochemical changes in chemical markers of sedimentary organic matter source and age, *Mar. Chem.*, *113*, 123-128.

Mayorga, E., Aufdenkampe, A. K., Masiello, C. A., Krusche, A. V., Hedges, J. L., Quay, P. D., Richey, J. E., and Brown, T. A.: Young organic matter as a source of carbon dioxide outgassing from Amazonian rivers, Nature 436, 538-541, 2005.

McCallister, S. L., & Giorgio, P. A. D. 2008. Direct measurement of the $\delta^{13}C$ signature of carbon respired by bacteria in lakes: Linkages to potential carbon sources, ecosystem baseline metabolism, and CO2 fluxes. *Limnol. Oceanogr. 53*(4), 1204.

McCallister, S. L., and del Giorgio, P. A. 2012. Evidence for the respiration of ancient terrestrial organic C in northern temperate lakes and streams. *PNAS*, *109*(42), 16963-16968.

McCallister, S. L., and del Giorgio, P. A. 2008. Direct measurement of the $\delta^{13}C$ signature of carbon respired by bacteria in lakes: Linkages to potential carbon sources, ecosystem baseline metabolism, and $CO_2$ fluxes, Limnol. Oceanogr. 53, 1204-1216.

McCallister, S. L., Bauer, J. E., Cherrier, J., and Ducklow, H. W. 2004. Assessing the sources and ages of organic matter supporting estuarine bacterial production: A novel multiple isotopes ($\delta^{13}C$, $\delta^{15}N$, $\Delta^{14}C$) approach. Limnol. Oceanogr. 49, 1687-1702, 2004.

McCallister, S. L., J. E. Bauer, H. W. Ducklow, E. A. Canuel. 2006. Sources of estuarine dissolved and particulate organic matter: A multi-tracer approach. *Org. Geochem*. 37, 454-468.

McCallister, S. L., J. E. Bauer, J. Cherrier, and H. W. Ducklow. 2004. Assessing the sources and ages of organic matter supporting estuarine bacterial production: A novel multiple isotopes ($\delta^{13}C$, $\delta^{15}N$, $\Delta^{14}C$) approach. Limnol. Oceanogr. **49:** 1687-1702.

McCarthy, M. D., Beaupre, S. R., Walker, B. D., Voparil, I., Guilderson , T. P., and Druffel, E. R. M.: Chemosynthetic origin of $^{14}C$-depleted dissolved organic matter in a ridge-flank hydrothermal system, *Nature Geoscience*, *4*, 32-36, 2011.

McCarthy, M., J. Hedges, and R. Benner. 1993. The chemical composition of dissolved organic matter in seawater. *Chem. Geol.* 107, 503-507.

McCarthy, M., J. Hedges, and R. Benner. 1996. Major biochemical composition of dissolved high molecular weight organic matter in seawater. *Marine Chemistry* 55, 281-297.

McCarthy, M., J. I. Hedges, R. Benner. 1998. Major bacterial contribution to marine dissolved organic nitrogen. *Science* 281, 231-234.

McCarthy, M., Pratum, T., Hedges, J., and Benner, R. 1997. Chemical composition of dissolved organic nitrogen in the ocean. *Nature 390*, 150-154.

McCaul, M. V., Sutton, D., Simpson, A. J., Spence, A., McNally, D. J., Moran, B.W., Goel, A., O'Connor, B., Hart, K., and Kelleher, B. P. 2011. Composition of dissolved organic matter within a lacustrine environment. Environ. Chem. 8, 146–154.

McConnaughey, T., and McRoy, C. P.: Food-web structure and the fractionation of carbon isotopes in the Bering Sea, Marine Biology 53, 257-262, 1979.

Mcknight, D.M., R. Harnish, R.L. Wershaw, J.S. Baron, and S. Schiff. 1997. Chemical Characteristics of Particulate, Colloidal, and Dissolved

Organic Matter in Loch Vale Watershed, Rocky Mountain National Park. *Biogeochemistry* 36, 99-124.

McManus, J., E. A. Heinen, and M. M. Baehr (2003), Hypolimnetic oxidation rates in Lake Superior: Role of dissolved organic material on the lake's carbon budget, *Limnol. Oceanogr.*, *48*, 1624-1632.

McNichol, A. P., and L. I. Aluwihare (2007), The power of radiocarbon in biogeochemical studies of the marine carbon cycle: Insights from studies of dissolved and particulate organic carbon (DOC and POC), *Chem. Rev.*, *107*, 443-466.

Meersche, K. V. D., Rijswijk, P. V., Soetaert, K., and Middelburg, J. J.: Autochthonous and allochthonous contributions to mesozooplankton diet in a tidal river and estuary: Integrating carbon isotope and fatty acid constraints, Limnol. Oceanogr. 54, 62-74, 2009.

Meili, M., Fry, B., and Kling, G. W.: Fractionation of stable isotopes ($^{13}$C, $^{15}$N) in the food web of a humic lake. Verh. Int. Ver. Limnol. 25, 501-505, 1993.

Meili, M., Kling, G. W., Fry, B., Bell, R. T., and Ahlgren, I.: Sources and Partitioning of organic matter in a pelagic microbial food web inferred from the isotopic composition ($\delta^{13}$C and $\delta^{15}$N) of zooplankton species. Arch Hydrobiol. Spec. Issues Adv. Limnol. 48, 53-61, 1996.

Meyer, J. L., R. T. Edwards, and R. Risley. 1987. Bacterial growth on dissolved organic carbon from a blackwater river. *Microb. Ecol.* 13, 13-29.

Meyers, P. A. 2003. Applications of organic geochemistry to paleolimnological reconstructions: a summary of examples from the Laurentian Great lakes. Org. Geochem. **34:** 261-289.

Meyers, P.A. 1997. Organic geochemical proxies of paleoceanographic, paleolimnologic, and paleoclimatic processes. *Org. Geochem.* 27, 213-250.

Miller, M. C., I. N. McCave, and P. D. Komar (1977), Threshold of sediment motion under unidirectional currents, *Sedimentology*, *24*, 507-527.

Miller, W.L. and R.G. Zepp. 1995. Photochemical production of dissolved inorganic carbon from terrestrial organic matter: significance to the oceanic organic carbon cycle. *Geophys. Res. Lett.* 22, 417-420.

Miller, W.L., M.A. Moran. 1997. Interaction of photochemical and microbial processes in the degradation of refractory dissolved organic matter from a coastal marine environment. *Limnol. Oceanogr.* 42, 1317-1324.

Minor, E. and Stephens, B. (2008) Dissolved organic matter characteristics within the Lake Superior watershed. *Organic Geochemistry* **39**, 1489-

1501.

Minor, E. C., B. J. Dalzell, A. Stubbins, K. Mopper. 2007. Evaluating the photoalteration of estuarine dissolved organic matter using direct temperature-resolved mass spectrometry and UV-visible spectroscopy. *Aquatic Sciences* 69, 440-455.

Minor, E. C., J. J. Boon, H. R. Harvey, A. Mannino. 2001. Estuarine organic matter decomposition as probed by direct temperature-resolved mass spectrometry and traditional geochemical technique. *Geochim. Cosmochim. Acta* 65, 2819-2834.

Minor, E., Boon, J., Harvey, H. and Mannino, A. (2001) Estuarine organic matter composition as probed by direct temperature-resolved mass spectrometry and traditional geochemical techniques. *Geochimica et Cosmochimica Acta* **65**, 2819-2834.

Minor, E.C., J. Pothen, B.J. Dalzell, H. Abdulla and K. Mopper. 2006. Effects of salinity changes on the photodegradation and UV-Visible absorbance of terrestrial dissolved organic matter. *Limnol. Oceanogr.* 51(5), 2181-2186.

Minor, E.C., J-P. Simjouw, J.J. Boon, A.E. Kerkhoff, and J. van der Horst. 2002. Estuarine/marine UDOM as characterized by size-exclusion chromatography and organic mass spectrometry. *Marine Chemistry* 78, 75-102.

Minor, E.C., J-P. Simjouw, J.J. Boon, A.E. Kerkhoff, and J. van der Horst. 2002. Estuarine/marine UDOM as characterized by size-exclusion chromatography and organic mass spectrometry. *Marine Chemistry* 78, 75-102.

Mohamed, M. N., and Taylor, W. D.: Relative contribution of autochthonous and allochthonous carbon to limnetic zooplankton: A new cross-system approach. Fund Appl. Limnol. 175, 113-124, 2009.

Monteith, T. J., and W. C. Sonzogni (1976), U.S. Great Lakes shoreline erosion loadings, Technical report of the international reference group of Great Lakes pollution from land use activities. IJC, Windsor, Ontario.

Moore, J. W., & Semmens, B. X. 2008. Incorporating uncertainty and prior information into stable isotope mixing models. *Ecology letters*, *11*(5), 470-480.

Mopper, K., A. Stubbins, J. D. Ritchie, H. M. Bialk, and P. G. Hatcher. 2007. Advanced instrumental approaches for characterization of marine dissolved organic matter: Extraction techniques, mass spectrometry, and nuclear magnetic resonance spectroscopy. *Chem. Rev.* 107, 419-442.

Moran, M.A. and R.G. Zepp. 1997. Role of photoreactions in the formation of biologically labile compounds from dissolved organic matter. *Limnol. Oceanogr*. 42, 1307-1316.

Moran, M.A., L.R. Pomeroy, E.S. Sheppard, L.P. Atkinson, R.E. Hodson. 1991. Distribution of terrestrially derived dissolved organic matter on the southeastern US continental shelf. *Limnol. Oceanogr.* 36, 1134-1149.

Mothersml, J. S. 1988. Paleomagnetic dating of late glacial and post-glacial sediments in Lake Superior, *Canadian Journal of Earth Sciences*, *25*, 1791-1799.

Munster, U., and R. J. Chrost. 1990. Origin, composition and microbial utilization of dissolved organic matter, p. 8-46. *In.* J. Overbeck and R. J. Chrost [eds.], Aquatic microbial ecology. Biochemical and molecular approaches. Springer.

Murphy, E. M., S. N. Davis, A. Long, D. Donahue and A. J. T. Jull (1985), $^{14}C$ in fractions of dissolved organic carbon in ground water, *Nature*, *337*, 153-155.

Nagao, S., T. Usui, M. Yamamoto, M. Minagawa, T. Iwatsuki, A. Noda. 2005. Combined use of $\Delta^{14}C$ and $\delta^{13}C$ values to trace transportation and deposition processes of terrestrial particulate organic matter in coastal marine environments. *Chemical Geology* 218, 63-72.

Nicholls, K. H. 1999. Effects of temperature and other factors on summer phosphorus in the inner bay of Quinte, Lake Ontario: implications for climate warming. J. Great Lakes Res. **25:** 250-262.

Nydal, R. 1963. Increase in radiocarbon from the most recent series of thermonuclear tests. *Nature* 200, 212-214.

Nydal, R., K. Lovseth, F. H. Skogseth. 1980. Transfer of bomb $^{14}C$ to the ocean surface. *Radiocarbon* 22, 626-635.

O'Loughlin, E. J., and Y.-P. Chin. 2004. Quantification and characterization of dissolved organic carbon and iron in sedimentary porewater from Green Bay, WI, USA. Biogeochemistry **71:** 371-386.

Ogawa, H., Amagai, Y., Koike, I., Kaiser, K. and Benner, R. (2001) Production of refractory dissolved organic matter by bacteria. *Science* **292**, 917-920.

Ogura, N. 1975. Further studies on decomposition of dissolved organic matter in coastal seawater. *Marine Biology* 31, 101-111.

Ogura, N. 1977. High molecular weight organic matter in seawater. *Marine Chemistry* 5, 535-549.

Opsahl, S. and R. Benner. 1997. Distribution and cycling of terrigenous dissolved organic matter in the ocean. *Nature* 386, 480-482.

Opsahl, S. and R. Benner. 1998. Photochemical reactivity of dissolved lignin in rivers and ocean waters. *Limnol. Oceanogr.* 43, 1297-1304.

Opsahl, S. P., and R. G. Zepp. 2001. Photochemically-induced alteration of stable carbon isotope ratios ($\delta^{13}C$) in terrigenous dissolved organic carbon. Geophys. Res. Lett. **28:** 2417-2420.

Ostrom, N. E., D. T. Long, E. M. Bell and T. Beals (1998), The origin and cycling of particulate and sedimentary organic matter and nitrate in Lake Superior, *Chemical Geology*, *152*, 13-28.

Ostrom, N. E., D. T. Long, E. M. Bell, and T. Beals. 1998. The origin and cycling of particulate and sedimentary organic matter and nitrate in Lake Superior. Chemical Geology **152:** 13-28.

Pace, M. L., Cole, J. J., Carpenter, S. R., Kitchell, J. F., Hodgson, J. R., Bogert, M. V., D. L., Bade, Kritzberg, E. S., and Bastviken, D.: Whole-lake carbon-13 additions reveal terrestrial support of aquatic food webs, Nature 427, 240-243, 2004.

Pang, P. C., and J. O. Nriagu (1977), Isotopic variations of the nitrogen in Lake Superior, *Geochim. Cosmochim. Acta*, *41*, 811–814.

Parnell, A. C., Phillips, D. L., Bearhop, S., Semmens, B. X., Ward, E. J., Moore, J. W., ... & Inger, R. 2013. Bayesian stable isotope mixing models. *Environmetrics*, *24*(6), 387-399.

Pearson, A., A. P. McNichol, B. C. Benitez-Nelson, J. M. Hayes, and T. I. Eglinton. 2001. Origins of lipid biomarkers in Santa Monica Basin surface sediment: A case study using compound-specific $\Delta^{14}C$ analysis. *Geochim. Cosmochim. Acta* 65, 3123-3137.

Pearson, A., T. I. Eglinton, and A. P. McNichol. 2000. An organic tracer for surface ocean radiocarbon. *Paleoceanography* 15, 541-550.

Pearson, K. (2005). An assessment of environmental changes during the Holocene as recorded in the bulk geochemical and molecular signals of lake superior sediments. M.S. Thesis, Water Resources Science, University of Minnesota, Duluth, MN, USA.

Pena, M. A., A. R. Lewis, and W. G. Harrison. 1991. Particulate organic matter and chlorophyll in the surface layer of the equatorial Pacific Ocean along 135° W. Mar. Ecol. Prog. Ser. **72:** 179-188.

Penner, J. E., H. Eddleman, and T. Novakov. 1993. Towards the development of a global inventory of black carbon emissions. Atmos. Environ. **27:** 1277-1295.

Peterson, B. and Fry, B. (1987) Stable Isotopes in Ecosystem Studies. *Annual Review of Ecology and Systematics* **18**, 293-320.

Peterson, B., B. Fry, M. Hullar, and S. Saupe. 1994. The distribution and stable isotopic composition of dissolved organic carbon in estuaries. Estuaries **17:** 111-121.

Petsch, S., Eglinton, T. and Edwards, K. (2001) $^{14}C$-dead living biomass: evidence for microbial assimilation of ancient organic carbon during shale weathering. *Science* **292**, 1127-1131.

Peuravouri, J. and K. Pihlaja. 1997. Molecular size distribution and spectroscopic properties of aquatic humic substances. *Anal. Chim. Acta* 337, 133-149.

Plante, A.F., Beaupré, S.R., Roberts, M.L. and Baisden, T. (2013) Distribution of radiocarbon ages in soil organic matter by thermal fractionation. *Radiocarbon* **55**, 1077-1083.

Pohlman, J. W., J. E. Bauer, W. F. Waite, C. L. Osburn, and N. R. Chapman (2011), Methane hydrate-bearing seeps as a source of aged dissolved organic carbon to the oceans, *Nature Geoscience*, *4*, 37-41.

Popp, B. N., E. A. Laws, R. R. Bidigare, J. E. Dore, K. L. Hanson, and S. G. Wakeham. 1998. Effect of phytoplankton cell geometry on carbon isotopic fractionation. Geochim. Cosmochim. Acta **62:** 69-77.

Prahl, F. G., J. R. Ertel, M. A. Goni, M. A. Sparrow, B. Evers-meyer. 1994. Terrestrial organic carbon contributions to sediments on the Washington margin. *Geochim. Cosmochim. Acta* 58, 3035-3048.

Pulido-Villena, E., Reche, I., Morales-Baquero, R.: Food web reliance on allochthonous carbon in two high mountain lakes with contrasting catchments: a stable isotope approach, Can. J. Fish. Aquat. Sci. 62, 2640-2648, 2005.

Quinn, F. H. (1992), Hydraulic residence times for the Laurentian Great Lakes, *J. Great Lakes Res.*, *18*, 22-28.

Rautio, M., Mariash, H., and Forsstrom, L.: Seasonal shifts between autochthonous and allochthonous carbon contributions to zooplankton diets in a subarctic lake, Limnol. Oceanogr. 56, 1513-1524, 2011.

Raymond, A. and J.E. Bauer. 2001. Use of $^{14}C$ and $^{13}C$ natural abundance for evaluating riverine, estuarine, and coastal DOC and POC sources and cycling: a review and synthesis. *Org. Geochem*. 32, 469-485.

Raymond, P. A. 2005. The composition and transport of organic carbon in rainfall: Insights from the natural ($^{13}C$ and $^{14}C$) isotopes of carbon, *Geophysical Research Letters*, *32*, L14402, doi: 10.1029/2005GL022879.

Raymond, P. A., and C. S. Hopkinson. 2003. Ecosystem modulation of dissolved carbon age in a temperate marsh-dominated estuary. Ecosystems **6:** 694-705.

Raymond, P. A., and J. E. Bauer (2001), Riverine export of aged terrestrial organic matter to the North Atlantic Ocean, *Nature*, *409*, 497-500.

Raymond, P. A., J. E. Bauer, N. F. Caraco, J. J. Cole, B. Longworth, and S. T. Petsch. 2004. Controls on the variability of organic matter and dissolved inorganic carbon ages in northeast US rivers. Marine Chemistry **92:** 353-366.

Raymond, P. A., & Cole, J. J. 2003. Increase in the export of alkalinity from North America's largest river. *Science*, *301*(5629), 88-91.

Reddy, C. M., A. Pearson, L. Xu, A. P. McNichol, B. A. Benner, S. A. Wise, G. A. Klouda, L. A. Currie, and T. I. Eglinton (2002), Radiocarbon as a tool to apportion the sources of polycyclic aromatic hydrocarbons and black carbon in environmental samples, *Environ. Sci. Technol.*, *36*, 1774-1782.

Redfield, A. C., B. H. Ketchum, and F. A. Richards. 1963. The influence of organisms on the composition of sea water, p. 26-77. *In* M. N. Hill [ed.], The sea. Wiley-Interscience.

Reimer, P. J. (2001), A new twist in the radiocarbon tale, *Science*, *294*, 2494-2495.

Repeta, D. J., and L. I. Aluwihare. 2006. Radiocarbon analysis of neutral sugars in high-molecular-weight dissolved organic carbon: Implications for organic carbon cycling. Limnol. Oceanogr. **51:** 1045-1053.

Repeta, D.J. and Aluwihare, L.I. (2006) Radiocarbon analysis of neutral sugars in high-molecular-weight dissolved organic carbon: Implications for organic carbon cycling. *Limnology and Oceanography* **51**, 1045-1053.

Repeta, D.J., T.M. Quan, L.I. Aluwihare, and A.M. Accardi. 2002. Chemical characterization of high molecular weight dissolved organic matter in fresh and marine waters. *Geochim. Cosmochim. Acta* 66, 955-962.

Richey, J.E., J. I. Hedges, A. H. Devol, P. D. Quay, R. Victoria, L. Martinelli, and B. C. Forsberg. 1990. Biogeochemistry of carbon in the Amazon River. *Limnol. Oceanogr*. 35, 352-371.

Roland, L. A., M. D. McCarthy, and T. Guilderson. 2008. Sources of molecularly uncharacterized organic carbon in sinking particles from three ocean basins: A coupled $\Delta^{14}C$ and $\delta^{13}C$ approach. Mar. Chem. **111:** 199-213.

Romankevich, E. A. 1984. Geochemistry of organic matter in the ocean. Springer.

Rose, K. A., E. L. Sikes, T. P. Guilderson, P. Shane, T. M. Hill, R. Zahn and H. J. Spero (2010), Upper-ocean-to-atmosphere radiocarbon offsets imply fast deglacial carbon dioxide release, *Nature*, *466*, 1093-1097.

Rosenheim, B.E., Day, M.B., Domack, E., Schrum, H., Benthien, A. and Hayes, J.M. (2008) Antarctic sediment chronology by programmed-temperature pyrolysis: Methodology and data treatment. *Geochemistry, Geophysics, Geosystems* 9, n/a-n/a.

Rosenheim, B.E., Roe, K.M., Roberts, B.J., Kolker, A.S., Allison, M.A. and Johannesson, K.H. (2013) River discharge influences on particulate organic carbon age structure in the Mississippi/Atchafalaya River System. *Global Biogeochemical Cycles* **27**, 154-166.

Roy, R., and G. Knowles. 1994. Effects of methane metabolism on nitrification and nitrous oxide production in polluted freshwater sediment. Appl. Environ. Microbiol. **60:** 3307-3314.

Russ, M. E., N. E. Ostrom, H. Gandhi, P. H. Ostrom, and N. R. Urban. 2004. Temporal and spatial variations in R:P ratios in Lake Superior, an oligotrophic freshwater environment. J. Geophys. Res. **109:** C10S12, doi: 10.1029/2003JC001890.

Russ, M. E., Ostrom, N. E., Gandhi, H., Ostrom, P. H., & Urban, N. R. 2004. Temporal and spatial variations in R: P ratios in Lake Superior, an oligotrophic freshwater environment. *Journal of Geophysical Research: Oceans*, *109*(C10).

Saliot, A., L. Mejanelle, P. Scribe, J. Fillaux, C. Pepe, A. Jabaud, and J. Dagaut. 2001. Particulate organic carbon, sterols, fatty acids and pigments in the Amazon River System. *Biogeochemistry* 53, 79-103.

Salonen, K., and Hammar, T.: On the importance of dissolved organic matter in the nutrition of zooplankton in some lake waters, Oecologia 68, 246-253, 1986.

Santos, G. M., J. R. Southon, K. C. Druffel-Rodriguez, S. Griffin, and M. Mazon. 2004. Magnesium perchlorate as an alternative water trap in AMS graphite sample preparation: A report on sample preparation at KCCAMS at the Univ. of California, Irvine. Radiocarbon **46:** 165-173.

Santschi, P. H., Balnois, E., Wilkinson, K. J., Zhang, J., Buffle, J., & Laodong, G. 1998. Fibrillar polysaccharides in marine macromolecular organic matter as imaged by atomic force microscopy and transmission electron microscopy. *Limnol. Oceanogr. 43*(5), 896-908.

Santschi, P. H., L. Guo, M. Baskaran, S. Trumbore, and J. Southon. 1995. Isotopic evidence for the contemporary origin of high molecular weight organic matter in oceanic environments. *Geochim. Cosmochim. Acta*. 59, 625-631.

Santschi, P.H., E. Balnois, E. Wilkinson, J. Buffle, L. Guo. 1998. Fibrillar polysaccharides in marine macromolecular organic matter as imaged by atomic force microscopy and transmission electron microscopy. *Limnol. Oceanogr*. 43, 896–908.

Saunders, G. 1976. Decomposition in fresh water, p. 341-374. *In*. J. Anderson, and A. MacFadyen [eds.], The role of terrestrial and aquatic organisms in decomposition processes. Blackwell.

Schallenberg, M., and C. W. Burns. 2004. Effects of sediment resuspension on phytoplankton production: teasing apart the influences of light, nutrients and algal entrainment, *Freshwater Biology*, *49*,143–159.

Schell, D. L.: Carbon-13 and Carbon-14 Abundances in Alaskan Aquatic Organisms: Delayed Production from Peat in Arctic Food Webs, Science 219, 1068-1071, 1983.

Schulz, K. L., & Sterner, R. W. 1999. Phytoplankton phosphorus limitation and food quality for Bosmina. *Limnology and Oceanography*, *44*(6), 1549-1556.

Schwede-Thomas, S.B., Y-P. Chin, K.J. Dria, P. Hatcher, E. Kaiser, and B. Sulzberger. 2005. Characterizing the properties of dissolved organic matter isolated by XAD and C-18 solid phase extraction and ultrafiltration. *Aquatic Science* 67, 61-71.

Semmens, B. X., Moore, J. W., & Ward, E. J. 2009. Improving Bayesian isotope mixing models: a response to Jackson et al.(2009). *Ecology letters*, *12*(3), E6-E8.

Shank, G. C., A. Evans, Y. Yamashita, and R. Jaffe .2011. Solar radiation-enhanced dissolution of particulate organic matter from coastal marine sediments, *Limnol. Oceonogr.*, *56*(2), 577-588.

Siebel, E., J. M. Armstrong, C. L. Alexander. 1976. Technical report on determination of quantity and quality of great lakes U.S. shoreline eroded material, Great Lakes Basin Commission, Ann Arbor.

Sierszen, M. E., G. S. Peterson, and J. V. Scharold. 2006. Depth-specific patterns in benthic-planktonic food web relationships in Lake Superior, *Canadian Journal of Fisheries and Aquatic Sciences*, *63*, 1496-1503.

Sigleo, A. C. 1996. Biochemical components in suspended particles and colloids: carbohydrates in the Potomac and Patuxent estuaries. *Org. Geochem*. 24, 83-93.

Sigleo, A. C., and J. C. Means. 1990. Organic and inorganic components in estuarine colloids: Implications for sorption and transport of pollutants. *Rev. Environ. Contam. Toxicol.* 112, 123-147.

Simjouw, J-P, E.C. Minor, and K. Mopper. 2005. Isolation and characterization of estuarine dissolved organic matter: Comparison of ultrafiltration and $C_{18}$ solid-phase extraction techniques. *Marine Chemistry* 96, 219-235.

Skoog, A., and R. Benner. 1997. Aldoses in various size fractions of marine organic matter: Implications for carbon cycling. *Limnol. Oceanogr.* 42, 1803-1813.

Smyntek, P. M., Teece, A. M., Schultz, K. L., Thackeray, S. J. 2007. A standard protocol for stable isotope analysis of zooplankton in aquatic food web research using mass balance correction models, Limnol. Oceanogr. 52, 2135-2146.

Sobczak, W. V., Cloern, J. E., Jassby, A. D., and Muller-Solger, A. B.: Bioavailability of organic matter in a highly disturbed estuary: the role of detrital and algal resources, PNAS 99, 8101-8105, 2002.

Sousounis, P. J., and E. K. Grover. 2002. Potential future weather patterns over the Great Lakes region. J. Great Lakes Res. **28:** 496-520.

Spencer, R. G., Hernes, P. J., Aufdenkampe, A. K., Baker, A., Gulliver, P., Stubbins, A., Aiken, G. R., Dyda R. Y., Butler, K. D., Mwamba, V. L., Mangangu, A. M., Wabakanghanzi, J. N., and Six, J. 2012. An initial investigation into the organic matter biogeochemistry of the Congo River. *Geochim. Cosmochim. Acta*, *84*, 614-627.

Stephens, B.M. and Minor, E.C. (2010) DOM characteristics along the continuum from river to receiving basin: a comparison of freshwater and saline transects. *Aquatic Sciences* **72**, 403-417.

Sterner, R. W. (2010), In situ-measured primary production in Lake Superior, *J. Great Lakes Res.*, *36*, 139-149.

Sterner, R. W., E. Anagnostou, S. Brovold, G. S. Bullerjahn, J. Finlay, S. Kumar, R. M. L. McKay and R. M. Sherrell. 2007. Increasing stoichiometric imbalance in North America's Largest Lake: nitrification in Lake Superior, *Geophys. Res. Lett.*, 34, L10406. doi:10.1029/2006GL028861.

Sterner, R. W., T. M. Smutka, R. M. L. McKay, Q. Xiaoming, and E. T. Brown. 2004. Phosphorus and trace metal limitation of algae and bacteria in Lake Superior. Limnol. Oceanogr. **49:** 495-507.

Stortz, K., R. Clapper, and M. Sydor (1976), Turbidity sources in Lake Superior, *J. Great Lakes Res.*, 2, 393-401.

Striegl, R. G., P. Kortelainen, J. P. Chanton, K. P. Wickland, G. C. Bugna, and M. Rantakari. 2001. Carbon dioxide partial pressure and $^{13}C$ content of north temperate and boreal lakes at spring ice melt. Limnol. Oceanogr. **46:** 941-945.

Strom, J. D. 2007. Seasonal evolution of dissolved gas concentration in western Lake Superior: Implications for respiration, primary production and carbon cycling, M.S Thesis, University of Minnesota.

Stubbins, A., Spencer, R. G., Chen, H., Hatcher, P. G., Mopper, K., Hernes, P. J., Mwanba, V. L., Mangangu, A. M., Wabakanghanzi, J.

N., and Six, J. 2010. Illuminated darkness: Molecular signatures of Congo River dissolved organic matter and its photochemical alteration as revealed by ultrahigh precision mass spectrometry. *Limnol. Oceanogr. 55*(4), 1467.

Stuiver, M. and Polach, H.A. 1977. Discussion; reporting of C-14 data. *Radiocarbon* **19**, 355-363.

Sundh, I. 1992. Biochemical composition of dissolved organic carbon derived from phytoplankton and used by heterotrophic bacteria. *Applied and Environmental Microbiology*, *58*(9), 2938-2947.

Syvaranta, J., and Rautio, M.: Zooplankton, Lipids and stable isotopes: importance of seasonal, latitudinal, and taxonomic differences, Can. J. Fish. Aquat. Sci. 67, 1721-1729, 2010.

Taipale, S., Kankaala, P., Tiirola, M., and Jones, R. I.: Whole-Lake dissolved inorganic $^{13}C$ additions reveal seasonal shifts in zooplankton diet, Ecology 89, 463-474, 2008.

Tambone, F., Genevini, P., D'Imporzano, G., and Adani, F. 2009. Assessing amendment properties of digestate by studying the organic matter composition and the degree of biological stability during the anaerobic digestion of the organic fraction of MSW. *Bioresource Technology*, *100*(12), 3140-3142.

Tanoue, E., Nishiyama, S., Kamo, M., and Tsugita, A. 1995. Bacterial membranes: possible source of a major dissolved protein in seawater. *Geochim. Cosmochim. Acta*, *59*(12), 2643-2648.

Tennant, C.J. (2016) Analysis of inorganic carbon and pH in the western arm of Lake Superior. Master's thesis, University of Minnesota, 93 p.

Thorp, J. H., and Delong, M. D.: Dominance of autochthonous autotrophic carbon in food webs of heterotrophic rivers? Oikos 96, 543-550, 2002.

Tranvik, L. J. 1990. Bacterioplankton growth on fractions of dissolved organic carbon of different molecular weights from humic and clear waters. *Appl. Environ. Microbiol.* 56, 1672-1677.

Tranvik, L. J.: Allochthonous dissolved organic matter as an energy source for pelagic bacteria and the concept of the microbial loop, Hydrobiologia 229, 107-114, 1992.

Tremblay L. and Benner R. 2009. Organic matter diagenesis and bacterial contributions to detrital carbon and nitrogen in the Amazon River system. *Limnol. Oceanogr.* **54**, 681-

Trumbore, S. 2009. Radiocarbon and soil carbon dynamics. Annu. Rev. Earth Planet. Sci. **37:** 47–66.

Trumbore, S. E. 1997. Potential response of soil organic carbon to global environmental change, *Proc. Natl. Acad. Sci.*, *94*, 8284-8291.

Turnbull, J. C., J. B. Miller, S. J. Lehman, P. P. Tans, R. J. Sparks, and J. Southon. 2006. Comparison of $^{14}CO_2$ , CO, and $SF_6$ as tracers for recently added fossil fuel $CO_2$ in the atmosphere and implications for biological $CO_2$ exchange. Geophys. Res. Lett. **33:** Lo1817, doi: 10.1029/2005GL024213

Urban, N. R., Auer, M. T., Green, S. A., Lu, X., Apul, D. S., Powell, K. D., and Bub, L. 2005. Carbon cycling in Lake Superior, J. Geophys. Res. 110, C06S90, doi:10.1029/2003JC002230.

Urban, N. R., J. Jeong, and Y. Chai. 2004*a*. The benthic nepheloid layer (BNL) north of the Keweenaw peninsula in Lake Superior: Composition, dynamics, and role in sediment transport. J. Great Lakes Res. **30:** 133-146.

Urban, N. R., Lu, X., Chai, Y., and Apul, D. S. 2004. Sediment trap studies in Lake Superior: Insights into resuspension, cross-margin transport, and carbon cycling, J. Gt. Lakes Res. 30, 147-161.

Vahatalo, A. V., and R. G. Wetzel. 2008. Long-term photochemical and microbial decomposition of wetland-derived dissolved organic matter with alteration of $^{13}C$ :$^{12}C$ mass ratio. Limnol. Oceanogr. **53:** 1387-1392.

Van Luven, D. M., J. E. Huntoon, and A. L. Maclean. 1999. Determination of the influence of wind on the Keweenaw Current in the Lake Superior basin as identified by advanced very high resolution radiometer (AVHRR) imagery, *J. Great Lakes Res.*, *25*, 625-641.

Velinsky, D. L. and M. L. Fogel . 1999. Cycling of dissolved and particulate nitrogen and carbon in the Framvaren Fjord, Norway: Stable isotopic variations, *Marine Chemistry*, *67*, 161-180.

Velinsky, D. L., and M. L. Fogel. 1999. Cycling of dissolved and particulate nitrogen and carbon in the Framvaren Fjord, Norway: Stable isotopic variations. Marine Chemistry **67:** 161-180.

Verdugo, P., A.L. Alldredge, F. Azam, D.L. Kirchman, U. Passow, and P.H. Santschi. 2004. The oceanic gel phase: a bridge in the DOM-POM continuum. *Marine Chemistry* 92, 67-85.

Verdugo, P., Alldredge, A. L., Azam, F., Kirchman, D. L., Passow, U., and Santschi, P. H. 2004. The oceanic gel phase: a bridge in the DOM–POM continuum. *Marine Chemistry*, *92*(1), 67-85.

Volkman, J.K., and E. Tanoue. 2002. Chemical and Biological Studies of Particulate Organic Matter in the Ocean. *J. Oceanogr.* 58, 265-279.

Wakeham, S. G., A. P. McNichol, J. E. Kostka, and T. K. Pease (2006), Natural-abundance radiocarbon as a tracer of assimilation of petroleum carbon by bacteria in salt marsh sediments. *Geochim. Cosmochim. Acta*, *70*, 1761-1771.

Wakeham, S. G., C. Lee, J. I. Hedges, P. J. Hernes, and M. L. Peterson. 1997. Molecular indicators of diagenetic status in marine organic matter. *Geochim. Cosmochim. Acta* 61, 5363-5369.

Wakeham, S. G., McNichol, A. P., Kostka, J. E., and Pease, T. K.: Natural-abundance radiocarbon as a tracer of assimilation of petroleum carbon by bacteria in salt marsh sediments, Geochim. Cosmochim. Acta 70, 1761-1771, 2006.

Walker, B. D., S. R. Beaupre, T. P. Guiderson, E. R. M. Druffel, M. D. McCarthy. 2011. Large-volume ultrafiltration for the study of radiocarbon signatures and size vs. age relationship in marine dissolved organic matter. *Geochim. Cosmochim. Acta* 75, 5187-5202.

Wang, X. C., E. R. M. Druffel, S. Griffin, C. Lee, M. Kashgarian. 1998. Radiocarbon studies of organic compound compound classes in plankton and sediment of the Northeast Pacific Ocean. *Geochim. Cosmochim. Acta* 62, 1365-1378.

Wang, X., R. F. Chen, and G.B. Gardner. 2004. Sources and transport of dissolved and particulate organic carbon in the Mississippi River estuary and adjacent coastal waters of the northern Gulf of Mexico. *Marine Chemistry* 89, 241-256.

Wang, X.-C., Callahan, J. and Chen, R.F. (2006) Variability in radiocarbon ages of biochemical compound classes of high molecular weight dissolved organic matter in estuaries. *Estuarine, Coastal and Shelf Science* **68**, 188-194.

Wang, X.-C., Druffel, E.R., Griffin, S., Lee, C. and Kashgarian, M. (1998) Radiocarbon studies of organic compound classes in plankton and sediment of the northeastern Pacific Ocean. *Geochimica et Cosmochimica Acta* **62**, 1365-1378.

Wattrus, N. J. and D. E. Rausch. 2001. A preliminary survey of relict shoreface-attached sand ridges in Western Lake Superior, *Marine Geology*, *179*, 163-177.

Weidel, B. C., Solomon, C. T., Kitchell, J. F., Cole, J. J., Coloso, J. J., Middelburg, J. J., ... & Carpenter, S. R. 1900. Does terrestrial organic carbon subsidize the planktonic food web in a clear-water lake?.

Weishaar, J.L., G.R. Aiken, B.A. Bergamaschi, M.S. Fram, R. Fugii, and K. Mopper. 2003. Evaluation of specific ultraviolet absorbance as an indicator of the chemical composition and reactivity of dissolved organic carbon. *Environ. Sci. Technol.* 37, 4702-4708.

Wershaw, R. L., J. A. Leenheer, and L. Cox. 2005. Characterization of dissolved and particulate natural organic matter (NOM) in Neversink Reservoir, New York. U.S. Geological Survey Scientific Investigations Report 2005-5108.

Willey, J. D., R. J. Kieber, M. S. Eyman, and G. B. Avery (2000), Rainwater dissolved organic carbon: concentrations and global flux, *Global Biogeochemical Cycles*, *14*, 139-148.

William, P.M., K.J. Robertson, A. Soutar, S.M. Griffin and E.R.M. Druffel. 1992. Isotopic signatures ($^{14}C$, $^{13}C$, $^{15}N$) as tracers of sources and cycling of soluble and particulate organic matter in the Santa Monica Basin, California. *Prog. Oceanog*. 30, 253-290.

Williams, P. M., Druffel, E. R. M., and Smith Jr, K. L.1987. Dietary carbon sources for deep-sea organisms as inferred from their organic radiocarbon activities, Deep-Sea Research 34, 253-266.

Wright, R., and J. Hobbie. 1965. The uptake of organic solutes in lake water. *Limnol. Oceanogr*. 10, 22-28.

Yurista, P. M., Kelly, J. R., and Miller, S. E.: Lake Superior zooplankton biomass: alternative estimate from a probability-based net survey and spatially extensive LOPC surveys, J. Great Lakes Res. 35, 337-346, 2009.

Zhu, J., C. Chen, E. Ralph, S. A. Green, J. W. Budd, and F. Y. Zhang. 2001. Prognostic modeling studies of the Keweenaw current in Lake Superior Part II Simulation, *J. Phys. Oceanogr.*, *31*, 396-410.

Zigah, P. K., E. C. Minor, and J. P. Werne. 2012. Radiocarbon and stable-isotope geochemistry of organic and inorganic carbon in Lake Superior. Global Biogeochem. Cycles, 26, *GB1023, doi:*10.1029/2011GB004132.

Zigah, P. K., E. C. Minor, J. P. Werne, and S. L. McCallister. 2011. Radiocarbon and stable carbon isotopic insights into provenance and cycling of carbon in Lake Superior. *Limnol. Oceanogr*. 56, 867-886.

Zigah, P. K., Minor, E. C., Werne, J. P., and McCallister, S. L. 2012. An isotopic ($^{14}C$, $\delta^{13}C$, and $\delta^{15}N$) investigation of the composition of particulate organic matter and zooplankton food sources in Lake Superior and across a size-gradient of aquatic systems. *Biogeosciences*, *9*, 3663-3678.

Zigah, P.K. (2012) Sources, biogeochemical cycling, and fate of organic matter in Lake Superior: An investigation using natural abundance radiocarbon and stable isotopes. University of Minnesota.

Zigah, P.K., Minor, E.C., Abdulla, H.A., Werne, J.P. and Hatcher, P.G. (2014) An investigation of size-fractionated organic matter from Lake Superior and a tributary stream using radiocarbon, stable isotopes and NMR. *Geochimica et Cosmochimica Acta* **127**, 264-284.

Zou, L., Wang, X.-C., Callahan, J., Culp, R.A., Chen, R.F., Altabet, M.A. and Sun, M.-Y. 2004. Bacterial roles in the formation of high-molecular-weight dissolved organic matter in estuarine and coastal waters: Evidence from lipids and the compound-specific isotopic ratios. *Limnology and Oceanography* **49**, 297-302.

Zou, L., X. Wang, J. Callahan, R. A. Culp, R. F. Chen, M. Altabet, M. Sun. 2004. Bacteria role on formation of high molecular weight dissolved organic matter in estuarine and coastal waters: evidence from lipids and the compound-specific isotopic ratios. *Limnol. Oceanogr.* 49, 297-302.